ANIMAL SCIENCE, ISSUES AND PROFESSIONS

MYCOTOXICOSES IN ANIMALS ECONOMICALLY IMPORTANT

Animal Science, Issues and Professions

Additional books in this series can be found on Nova's website under the Series tab.

Additional e-books in this series can be found on Nova's website under the E-books tab.

ANIMAL SCIENCE, ISSUES AND PROFESSIONS

MYCOTOXICOSES IN ANIMALS ECONOMICALLY IMPORTANT

EDLAYNE GONÇALEZ
JOANA D'ARC FELICIO
AND
SIMONE AQUINO
EDITORS

Nova
Nova Science Publishers, Inc.
New York

For permission to use material from this book please contact us:
Telephone 631-231-7269; Fax 631-231-8175
Web Site: http://www.novapublishers.com

LIBRARY OF CONGRESS CATALOGING-IN-PUBLICATION DATA

Mycotoxicoses in animals economically important / editors, Edlayne Gongalez,
Joana D'Arc Felicio and Simone Aquino.
p. cm.
Includes bibliographical references and index.
ISBN 978-1-61668-195-1 (hardcover)
1. Mycotoxicoses in animals. 2. Livestock--Diseases. 3. Veterinary
mycology. 4. Mycotoxins. I. Gongalez, Edlayne. II. Felicio, Joana D'Arc.
III. Aquino, Simone.
[DNLM: 1. Animals, Domestic. 2. Mycotoxins--toxicity. 3. Veterinary
Medicine. SF 757.5 M995 2010]
SF780.7.M93 2010
636.089'595295--dc22
2010016717

Published by Nova Science Publishers, Inc. ✝ New York

Contents

FOREWORD

"MYCOTOXICOSES IN ANIMALS ECONOMICALLY IMPORTANT"

Health effects in animals or humans from fungal contamination of feeds or foods have been recognized for hundreds of years. Notable examples in veterinary medicine were ergotism of livestock and moldy corn poisoning of equines. Both of these were recognized as clinical entities long before the toxic agents were identified. The modern age of mycotoxin study advanced dramatically after a 1960 incident known as "Turkey X disease" where more than 100,000 turkeys died in England from consumption of imported peanut meal. Extensive and intensive studies revealed similar signs and lesions on other animals, and led to the recognition 1961 that the diseases were a result of certain *Aspergillus* spp. that produced a toxin which we now recognize as aflatoxin. As a veterinary student in the early 1960s, I was aware of the name "aflatoxin" but relatively little was known of its multisystemic implications and eventual world- wide impact leading even to recognition as a food based hepatocarcinogen. Compare this to today's knowledge and literature with thousands of publications on aflatoxin ranging from its cause and occurrence to the molecular and genetic basis for its effects. And yet today we still struggle to control its occurrence and to reduce its effects in our animal populations around the world.

With the recognition of aflatoxin as a significant human and animal threat to health and productivity we continue to note that more than 300 mycotoxins are recognized, but hard evidence for substantial damage to health or productivity has been gained for only 10 to 15 mycotoxins and is centered on six to ten prominent toxins from several genera of molds (*Aspergillus, Penicillium, Fusarium*). The past fifty years have seen massive improvements in detailed knowledge of these major mycotoxins, from the conditions for their formation to their acute effects and finally to some understanding of their subtle or interactive effects in production agriculture. Advanced research efforts and greatly improved storage and preventive measures have improved our knowledge and control of mycotoxin threats. However, much of the world is still quite susceptible to mycotoxin contamination because mycotoxin formation is highly weather dependent and influenced by storage limitations.

In animal production, dramatic acute mycotoxin poisoning with severe or fatal disease is of relatively low prevalence. However, uncertainty and controversy continue to plague livestock producers and health professionals based on real or potential damage to feed efficiency, immune system function, inefficient animal reproduction and possible residues in the human food supply. Suspicion of mycotoxin related disease continues, often without adequate tools to fully assess the threat on individual farms – resulting in economic decisions being made based on insufficient available information or on reaction with regard to current best practices and evidence based medicine and husbandry. The greatest challenges and continuing costs for the livestock industry is the prediction and detection of relevant mycotoxins in the feed supply, accurate detection of concentrations of known mycotoxins in feedstuffs, and better evaluation of subtle mycotoxin effects on health and productivity. Answers to these and other questions will come not only from applicable research, but from better field testing for mycotoxins and careful clinical evaluation of scenarios where moldy feed or mycotoxins can be strongly linked to animal disease.

This new volume of mycotoxin reviews for major livestock species is an important step in providing an international perspective on mycotoxins, especially important with the huge volume of international grain trade. It should be useful to livestock producers, veterinarians and nutritionists who work with known mycotoxin contamination or the more difficult area of suspected mycotoxin issues that must be distinguished from other animal health problems. The book provides documentation of current knowledge as well as perspective on some unresolved issues and problems that continue to need the attention of researchers, regulators and animal health professionals to assist in detection and control of mycotoxin effects. Information covered ranges from challenges to improved analytical efforts, improved prevention of contamination and poisoning, better understanding of mycotoxin interactions with nutrients or other mycotoxins and continued increase in research on the nutritional, immunological and reproductive effects of our well known mycotoxins. Beyond this, the writers engage readers to consider all aspects of mycotoxin formation and effects and to look toward new evidence of effects from emerging mycotoxins. Only by reviewing what we already know and continuing to question where we should be placing new effort will the mycotoxin challenge be met. This volume is a positive step in those efforts.

Gary Osweiler, DVM, PhD
Professor, Veterinary Diagnostic and Production Animal Medicine

PREFACE

Mycotoxins are secondary metabolites produced by specific filamentous fungi that infest agricultural commodities. They are known as a contaminant crops and consequently animal feeds and animal products, causing significant economic losses associated with their impact on, animal health and productivity in agribusiness. A same fungus is able to produce more than one kind of mycotoxins that are not visible, odor and taste. The impact of mycotoxins on animal performance is a complex topic because the animal symptoms vary according to the animal species, sex, age, breed, general health, immune status, nutritional standing and in their susceptibility to specifics mycotoxins. Besides, moulds and mycotoxin contamination may occur in any point of the supply chain and it represents a challenge to different experts.

Avoiding fungi contamination on the field is difficult or impossible, considering many factors as high crop densities, insect or mechanical damage and late harvesting. The main interventions that may be employed for the control of mycotoxins are prevention of contamination, identification and segregation of contaminated material, and detoxification.

The steps of prevention that are the most effective are those carried out before the fungal infestation and before mycotoxin production occur on plant material. Several agricultural practices can influence the crop contamination as preventing it during storage can be done to some extend with proper storage conditions, controlled temperature and moisture content and good management practices, but even the best management of agricultural strategies cannot totally eradicate mycotoxin contamination. Many physical, chemical or biological treatments have been tested against toxigenic fungus *in vitro*, in the greenhouse or under field conditions.

This book will describe conditions that favor mycotoxin formation and consumption, suggesting preventive approaches. The authors discussed in details the clinical symptoms and economic impacts of mycotoxicosis in horses, ruminants, swine and domestic fowls. In addition, the treatment, control methods and, synergistic effects of different mycotoxins in the food was reported under a different point of view of specialists.

In view of the large economic losses and problems of animal health caused by contamination of food by mycotoxins, we decided to discuss the consequences of ingestion of mycotoxins by animals of economic importance, due to the absence of documents that brings together all the information.

We expect that this book will be useful for all professionals and students interested in the mycotoxins area.

E. Gonçalez, S. Aquino and J.D. Felicio

In: Mycotoxicoses in Animals Economically Important ISBN: 978-1-61668-195-1
Editors: Edlyane Gonçalez, et. al. pp. 1-34

Chapter 1

EQUINE MYCOTOXICOSIS

Gary D. Osweiler *
Professor of Veterinary Toxicology,
Department of Veterinary Diagnostic and Production Animal Medicine,
College of Veterinary Medicine, Iowa State University, Ames, Iowa

ABSTRACT

Molds and mycotoxins are of worldwide concern for health and productivity of animals. Horses utilize grains and forages that may contain mycotoxins [1-3]. These toxins cause clinical and economic problems of food and fiber animals worldwide. Sometimes they are considered as a group with similar clinical effects; but, in reality the clinical effects of mycotoxins differ widely, with quite different effects in many major body systems and highly variable animal susceptibility related to age, species, sex and production systems. Thus, these widely different chemical structures have but one basic similarity - they are chemicals produced by mold growth on or in susceptible substrates (grains, forages or other organic matter). Successful management and prevention of mycotoxicoses requires knowledge of their formation and control in addition to knowledge of clinical history, consideration of differential diagnoses, and often reliance on clinical laboratory and analytical chemistry assays for diagnostic confirmation. This review will document and describe conditions that favor mycotoxin formation and consumption, suggest preventive approaches, characterize the clinical and economic effects of mycotoxins for horses, recommend diagnostic confirmation and outline therapy, supportive care and remediation for major mycotoxicoses. Mycotoxins of major

* Corresponding author for proof and reprints:
Gary D. Osweiler, DVM, PhD
1583 Veterinary Medicine
Department of Veterinary Diagnostic and Production Animal Medicine
College of Veterinary Medicine
1600 South 16th Street
Iowa State University
Ames, Iowa 50011
Phone: (515) 294-1950
Fax: (515) 294-3564
Email: osweiler@iastate.edu

health or economic importance for horses include aflatoxins, fumonisins and ergopeptide alkaloids. Mycotoxins of concern on an occasional or regional basis vary widely but involve ryegrass tremorgens, paspalum tremorgens, slaframine and satratoxins. Additional mycotoxins are of potential importance because of their occurrence regionally or worldwide, but there is little evidence of their effects on horses and limited or no direct experimental documentation of their effects. This chapter will cover established and recent information for the major mycotoxins and also review briefly the expected issues and problems for less common mycotoxins for the horse. In addition, general principles of mycotoxin formation, risk, diagnosis and control will be summarized.

INTRODUCTION

Conditions that favor mold growth on various commodities will promote mycotoxin production [4]. Mycotoxins are considered as secondary metabolites of mold growth. A secondary metabolite is a product of fungal metabolism that is not essential for growth or reproduction, but may in some way enhance survival of the fungus (Figure 1). Evidence of the antibiotic effect of many fungal metabolites has resulted in development of early as well as modern antibacterial compounds by control and manipulation of fungal growth conditions. These secondary metabolites are produced in response to plant injury or fungal stress, and the resulting mycotoxins can range from acute toxicants to subtle inhibitors of growth or reproduction in animals. Thus, common risk factors for mycotoxins are the circumstances that favor their formation and persistence in feedstuffs consumed by livestock including horses. Understanding these factors, provides a somewhat common pathway to preventing many of their effects by reducing their formation or their persistence in feeds and forages utilized by horses.

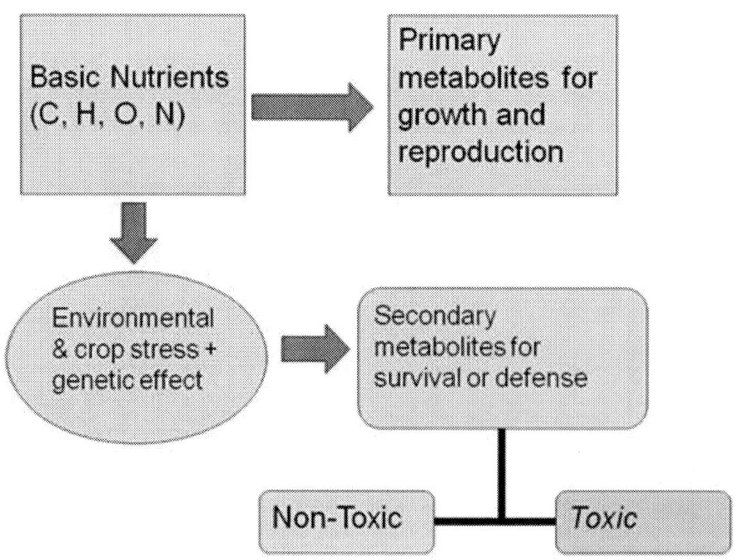

Figure 1. Mycotoxin Formation by Fungi occurs when fungal stress stimulates secondary metabolites, some of which may be toxic (i.e. mycotoxins).

Some 300 – 500 estimated mycotoxins are known to occur either in nature or by laboratory manipulation. At present, worldwide recognition and practical knowledge of important clinical conditions is limited to perhaps less than 30 mycotoxins, and extensive knowledge of formation, toxicity, mechanism of action, effects and control relevant to farm animals and horses is limited to perhaps 20 toxins or toxin groups. This chapter will focus on the mycotoxins of demonstrated clinical importance to horses.

SOURCES AND OCCURRENCE OF MYCOTOXINS

Mold Growth and Mycotoxin Production in Crops

Understanding the conditions that promote mycotoxin formation and control should be part of the comprehensive management of mycotoxin risk. A veterinarian may be the most frequent professional visitor to equine farms and stables, and a general knowledge of the sources and conditions that lead to fungal growth and mycotoxin formation can help owners and caretakers prevent or minimize feed contamination and the health effects that follow. Veterinarians should know basic principles, sources and conditions of mycotoxin risk and enlist agronomists, plant pathologists and chemical analysts as needed in the prevention and control of mycotoxin risk. Detailed information on mycotoxin formation and occurrence is well-documented and selected information to recognize high-risk circumstances are featured in this chapter [5, 6, 7, 8, 9].

Mycotoxin synthesis depends on mold growth, but mold growth does not always produce mycotoxins [8, 10]. Mycotoxin formation can result from stresses or limitations to fungal growth in response to adverse plant and environmental factors. Environmental conditions that favor mold growth and mycotoxin production often include seed coat injury from drought, insect damage or mechanical harvest as well as temperature and moisture conditions that favor mycotoxin synthesis.[5, 6, 8,10] Both feed grains (e.g. corn, wheat, barley, milo, cottonseed) and forages (fescue, ryegrass, red clover, dallis grass) support fungal growth that leads to mycotoxin production. Fungi can be external on grain or invade grains as a result of seed coat damage, and the seed coat is a major factor in protection from fungal invasion and potential mycotoxin synthesis. Fungal growth requires available energy (carbohydrate, usually as starch or cellulose), moisture, oxygen and suitable temperatures, often 12-37 $^{\circ}$C [8, 9]. All four of these factors are essential for mold growth which can lead to toxin production. General conditions that favor selected mycotoxin producing fungi are compared in Table 1. Notice that optimal temperature and moisture ranges vary substantially among different fungi. Because mycotoxin production is affected by fungal growth in conjunction with weather and crop conditions, local authorities and technical specialists can often recommend when conditions favor fungal infestations that are likely to lead to mycotoxin contamination.

Table 1. Substrates and conditions favoring formation of selected mycotoxins of risk to horses

MYCOTOXIN	Fungi and Substrates	Moisture	Temperature
Aflatoxins B1, B2, G1, G2*	*Aspergillus flavus, A. parasiticus & A. nominus* in corn, milo, peanuts, cottonseed	ERH 80-85%* EMC 13-32%**	24-32°C including high night time temperatures, enhanced by drought and insect stress
Deoxynivalenol	*Fusarium roseum* on corn, wheat	EMC 24-26%	< 21 ^{O} C
Ergopeptide alkaloids	*Claviceps purpurea* in wheat, barley, rye, oats, triticale, grasses (ergotamine, ergocornine, ergocristine, ergometrine, ergovaline) *Neotyphodium coeniphialum* endophyte in Tall Fescue (*Festuca arundinacea)* producing Ergovaline	Moist conditions during seed formation; spread is enhanced by wind and insects. Probable EMC > 20% Nitrogen fertilization, overgrowth or stockpiling favor endophyte infection and transmission	Cool to warm temperate conditions
Fumonisins	*Fusarium verticilloides* in corn	Moderate drought stress in corn during silking or ear development	Moist, humid conditions during crop maturation and drying
Slaframine	*Rhizoctonia leguminicola*(Black Patch Disease) on Clovers (*Trifolium sp*)	Warm to cool and moist conditions favor Black Patch disease with production of slaframine	Ambient temperatures approximately 20 – 30 ^{O} C. Preformed toxin stable in hay
Tremorgens Lolitrems Paspalitrem	Endophyte fungus (*Neotyphidium lolii*) in perennial ryegrass (*Lolium perenne*) *Claviceps paspali* ergot on Dallis Grass (*Paspalum dilatatum*)	Moist conditions during seed formation support infection and spread of fungus	Warm to cool temperatures during growth and seed formation

*ERH = Equilibrium Relative Humidity EMC = Equilibrium Moisture Concentration**

Conditions are approximations for environmental conditions and can vary from stated values in the description.

Detection of Mycotoxins in Contaminated Grain

Although mold growth is necessary for mycotoxin synthesis, simple visual or cultural examination of grain or feed is not adequate to predict mycotoxin contamination. There is limited to no correlation between mold spore counts or amount of fungal growth and contamination by mycotoxins. Plant or fungal stress influences generation of a mycotoxin. Nor does lack of mold growth assure that a feed or forage is free from mycotoxins. Very low grain moisture content, high temperature drying, high temperature and pressure during milling, or chemical treatments (e.g. propionic acid mold inhibitors) may reduce or prevent mold growth even though previous activity had produced a mycotoxin [8, 9]. Mycotoxins can persist in mold-free feeds because most are resistant to drying and the high temperatures and pressures of processing that may destroy or control fungi but leave the mycotoxin intact [10].

Field and Storage Fungi

Mycotoxin can be formed either in crops or in storage after harvest. Fungi have been classified as field fungi or storage fungi [8, 9]. Field fungi develop prior to harvest, and *Fusarium* spp. are major field fungi favored by high relative humidity (>70%) and grain moisture (>22%) for growth. They grow poorly when stored grain is < 22% moisture, and toxin production does not occur readily if dry grain is remoistened [8, 9]. Fungal infection damages developing seeds, with shriveling and weakening or death of embryos. Common names for this effect are "weathering" or “tombstone” kernels recognized at harvest time. These quality effects may be consistent with increased risk of storage fungi and their related mycotoxins (trichothecenes, deoxynivalenol and zearalenone). Fungal genera known as storage fungi include *Aspergillus* spp. and *Penicillium* spp. which are responsible for production of aflatoxins, ochratoxins and citrinin. Storage fungi can grow and produce mycotoxins at relatively low grain moisture concentrations of 15 to 18% and within a temperature range from 10 to 50°C. However, “storage fungi” may be a misleading term, since *Aspergillus flavus* often produces significant concentrations of aflatoxins in crops prior to harvest, especially when conditions of drought and insect damage occur during grain formation and maturation [8, 9]. Storage conditions will substantially enhance mold growth and toxin production from storage fungi. Grain stored at higher than acceptable moisture for storage may continue to respire and produce water. This can lead to accumulation of free-moisture levels, known as *water activity* that supports mold growth and toxin production. Water activity (*Aw*) is the ratio of the vapor pressure of water in the grain to that of pure water at the same temperature and pressure and these are expressed as a ratio less than 1.0. A water activity from 0.70 to 0.90 typically will support mold growth in stored grains. Alternating warm and cool seasonal temperatures with diurnal variation in temperatures promotes condensation and moisture migration within a storage bin and increased *Aw* in localized areas that favor mold growth. When fungal-contaminated grain is cracked or ground the protective seed coat is broken allowing fungal access to the carbohydrate rich endosperm and favoring mold growth. Feed stored in warm, humid conditions may mold and produce mycotoxins within only a few days [6, 7, 11]. Aflatoxin-producing *Aspergillus* spp. have been demonstrated to produce substantial increases of aflatoxin in stored grain within several (2 – 7) days in storage.

Cleaning and storage practices can influence mycotoxin contamination of grains and finished concentrate feeds. Mycotoxins are often increased in damaged, light or broken grain. Feeding poor quality grain or screenings (high concentration of defective grain) increases the probability of mycotoxin contamination [8, 9, 12]. Cleaning of damaged grain almost always reduces the concentration of mycotoxin remaining in the retained grain. For those mycotoxins produced in forages, their position as endophytes (within the seed, leaf or stalk) precludes physical separation of contaminated portions of the forage, so cultural management of the host forages is necessary to prevent contamination.

Effects of Commerce and Movement of Grains

Mycotoxin production may be sporadic by season, geographic region, drought and insect damage and prevailing harvest and storage practices. Certain geographic regions are considered at high risk for specific mycotoxins [6, 8, 13]. Regional and seasonal effects are often obscured by transport or blending of grains, damage in transit, and poor storage conditions [10]. Production history for a suspect grain is an important clue in investigation of a potential mycotoxicosis, but is difficult when massive blending and high volume transportation is utilized for distribution. Grain collection and purchase points, including export sites, often only check for aflatoxins and primarily in years when growing conditions suggest a potential problem. A higher level of inspection and testing may be required by regulatory requirements of individual countries or specific grain marketing companies or organizations. Grain sold locally and not entering interstate commerce or international trade may never be tested or characterized other than by visual observation and grading.

Recently a wide range of rapid kit-based mycotoxin testing procedures are available to evaluate mycotoxins in grains, and less commonly in forages. These methods can be highly effective if used as directed by the manufacturer. Their reliability is often confirmed by specific government agencies. For example, in the USA, mycotoxin test kits are tested and approved under the Grain Inspection, Packers and Stockyards Act (GIPSA) which adds a third party assurance about the reliability of a test method and identifies the commodities for which that specific method has been effective. Occasionally if a specific kit-based test is used for commodities not evaluated by GIPSA (for example a method approved for corn is used on silage or alfalfa) the reliability for that commodity should be questioned.

Risk Factors for Occurrence of Specific Mycotoxins

Relatively few specific studies of comparative mycotoxin prevalence are available. Mycotoxin formation and prevalence of toxicosis varies widely in the various climates and agriculture around the globe. Important grain mycotoxins worldwide with current relevance to horses appear to include aflatoxins, fumonisins, deoxynivalenol (vomitoxin) and ergot [3, 5, 12]. Forage-related mycotoxins important to horses include slaframine and swainsonine from red clover (*Trifolium praetense*), lolitrems from perennial ryegrass (*Neotyphodium lolii*) and ergopeptide alkaloids from Tall Fescue (*Neotyphodium coenophialum*) and ergot sclerotia (*Claviceps purpurea*) [1, 3, 6].

SOURCES AND OCCURRENCE OF SPECIFIC MYCOTOXINS

Management of mycotoxin risk is enabled by knowledge of specific conditions and characteristics of mold infections and mycotoxin productions. This section provides some specific information and examples of conditions that favor contamination of grain or forages by selected common mycotoxins.

Aflatoxins

Aspergillus flavus, *A. parasiticus* and less commonly *A. nominus* produce aflatoxins [6, 7, 10, 12, 14]. They are associated primarily with warm humid regions of the southeastern USA, as well as limited areas of the southwestern USA and in temperate to tropical regions worldwide. Grains frequently contaminated with aflatoxin include corn, milo, sorghum, cotton and peanuts. Seed coat damage (insects, drought stress and mechanical harvest) highly favors fungal invasion and toxin production [7, 8]. Grain is the preferred substrate, and aflatoxin is rarely formed in or on forages or vegetative growth of grain crops. Local conditions such as irrigation of intensively grown cotton in desert areas can also favor aflatoxin production. Under favorable conditions aflatoxins are produced rapidly, with from 950 – 2800 parts per billion (ppb) present 4 days post inoculation in field corn. The *A. flavus* infection also reduces the starch content of corn by 12%, reducing the overall energy value of infected seeds [15]. Endemic regions may have a continuing prevalence of aflatoxin in most years, due to consistent environmental and cultural stress. These stresses in conjunction with high ambient day and night temperatures above $21^{\circ}C$ ($70^{\circ}F$) and high relative humidity promote mold invasion and aflatoxin production. Drought and insect damage in non-endemic areas can lead to mold infestation with aflatoxin contamination, and pre-harvest crop condition reports are useful early indicators of potential aflatoxin contamination [8, 12].

Aflatoxins B_1, B_2, G_1 and G_2 and other minor aflatoxins occur in grains. Aflatoxin B_1 is the principal and most toxic fraction produced under natural conditions [5, 16, 17]. Purified aflatoxins have been considered by some as less toxic than the mixture of toxins associated with natural production in grains, but experimental confirmation and quantification of this difference is scarce and inconsistent [18]. Most large animal studies have been done with culture material or naturally contaminated grains. Aflatoxin metabolites are partially excreted in milk or urine primarily as the metabolite aflatoxin M_1, and in milk they can serve as a source of toxin for highly susceptible neonatal nursing animals [19].

Fumonisins

Fusarium verticilloides and *F. proliferatum* are recognized worldwide as producers of the fumonisin mycotoxins [5, 20-24]. They are important because of their common association with corn and animal mycotoxicoses [25]. Fumonisins are recognized as the cause of equine leukoencephalomalacia (ELEM), a major equine mycotoxin identified only recently [26]. Corn is the commodity almost exclusively associated with clinically relevant fumonisin contamination, and both white and yellow corn is affected. Insect damage and adverse weather conditions are associated with increased fumonisin production, and fumonisins are at highest concentration in broken kernels and screenings [6, 20, 24, 27]. Screenings and other seriously damaged kernels are the major source of fumonisins in a corn supply, and removal of corn screenings quite substantially reduces fumonisin contamination [28]. Conditions favoring fumonisin production appear to include a period of drought during the growing season with subsequent cool, moist conditions during pollination and kernel formation. Low concentrations are commonly found in corn in many parts of the world [12, 23].

Three principle fumonisins of toxicological importance in corn are designated as Fumonisin B_1 (FB_1), Fumonisin B_2 (FB_2) and Fumonisin B_3 (FB_3) [2, 3, 6, 20, 22, 24]. Naturally-contaminated corn contains approximately three times more FB_1 than FB_2 and they are nearly equivalent in toxicity, while FB_3 is much less toxic [29]. Fumonisins are an aliphatic hydrocarbon with a terminal amine group and two tricarboxylic acid side chains; FB_1, FB_2 and FB_3 are designated according to the number and position of hydroxyl groups. Unlike other common mycotoxins, fumonisins are water soluble as well as heat stable and resistant to alkali treatments that inactivate many other common mycotoxins [6, 21].

Trichothecene Mycotoxins

Trichothecenes, including T-2 toxin, diacetoxyscirpenol, deoxynivalenol (vomitoxin) and other minor or less common trichothecenes are produced by *Fusarium sporotrichioides* and *F. graminearum* in many temperate climates of the world (North America, Northern Europe, Russia, China). Cool moist conditions favor a high prevalence of *Fusarium* spp. associated with trichothecene production compared to tropical or subtropical climates. Cool moist conditions with alternating warm and cool temperatures typically favor toxin production [8, 30]. Corn, wheat and other cereal grains are common commodities susceptible to trichothecene production [7, 8]. However, Deoxynivalenol (DON) is the principal trichothecene routinely documented to occur at clinically relevant concentrations. Claims for occurrence of other trichothecenes (T-2 toxin, diacetoxyscirpenol, nivalenol, neosolaniol) are infrequent and/or usually occur at low concentrations in harvested grain [31].

The trichothecenes are approximately 150 structurally related compounds with toxic effects that arise from a 12, 13-epoxide in the sesquiterpene ring. Four groups are designated as A, B, C and D based on chemical structure. T-2 toxin and diacetoxyscirpenol (DAS) are in Group A. Group B is primarily comprised of nivalenol and deoxynivalenol (DON, vomitoxin).

Trichothecene toxins are stable in the environment, and resistant to the heat and pressure of cooking, feed milling and processing [32]. DON is stable to heat at acid pH (4.0) but partially destroyed by $100^{\circ}C$ heat at pH 10 and totally destroyed at $120^{\circ}C$ and pH 10 [33].

The group B trichothecene deoxynivalenol (DON, trivial name vomitoxin) is considered less toxic than T-2 and DAS, but is highly prevalent in feed grains of temperate regions and well documented as a cause of feed refusal in domestic livestock.

Stachybotryotoxins (Satratoxins)

Stachybotrys atra is a filamentous mold that favors growth on a cellulose-rich substrate. Most frequently identified energy sources have included forages, coarse grains, paper and wood [34]. The toxin is a macrocyclic trichothecene which includes several fractions designated as F, G and H. Reports from some parts of the world have implicated this group commonly known as satratoxins in toxicosis of sheep, cattle, swine, poultry and horses. An outbreak in Eastern Europe with death in horses was attributed to *Stachybotrys* [35]. Barley straw was contaminated with a blackened fungus identified as *Stachybotrys alternans*. Feeding infected straw containing the fungus to experimental horses reproduced the newly identified disease, which was named "stachybotryotoxicosis". More recently, there have been reports of numerous mycotoxins isolated from *Stachybotrys atra* from various parts of Hungary and Czechoslovakia. Rice cultures analyzed by high-performance liquid chromatography revealed 17 isolates, of which 14 produced macrocyclic trichothecenes (satratoxins F, G, and H, roridin E, and verrucarin J) as well as trichoverrols A and B. Satratoxins G and H were present in higher quantities than other trichothecenes [36]. Today there are occasional occurrences of disease clinically suggestive of satratoxicosis, but generally limited evidence to document satratoxin toxicosis in horses, and it remains virtually unknown as a clinical entity in North America [3, 20]. The association of *Stachybotrys* with "Sick Building Syndrome" and related human health issues keeps this mold and its associated mycotoxins as a continued point of interest for public health as well as veterinary medicine.

Ergot Alkaloid Mycotoxins from *Claviceps* spp.

Two species of ergot fungi continue to be important to animal agriculture throughout the world. They are *Claviceps purpurea* and *Claviceps paspali* [12]. The former is undoubtedly most important of the two because it commonly invades both pasture grasses used for grazing and hay as well as small grains used in concentrate feeds. Commodities that include wheat, barley, oats, rye, and triticale are affected. Ergotism was a mycotoxicoses putatively a cause of human poisoning in the middle ages. In modern times it has been associated with important livestock damage including dry gangrene of the extremities (feet, ears, tail) in cattle, swine and sheep. Its role in peripheral vascular disease of horses appears minor. However, it is, along with Tall Fescue endophytes a high risk issue for equine reproduction and foal survival and will be discussed in detail later as a clinical entity.

An ergot sclerotium (ergot, or ergot body) is produced by *C. purpurea* by invasion of the ovary of numerous grasses and small grains (tall fescue, bromegrass, orchard grass, bluegrass; wheat, barley, oats, rye, triticale). The life cycle of the ergot body begins with ascospores spread by wind to serve as a primary infective source. Infection of the ovary of the grass flower forms a sclerotium (ergot) that is usually slightly larger and differently shaped than the grass or small grain seed [6,10]. The sclerotia of *C. purpurea* are elongated and slightly curved, somewhat larger than the original grass ovary and with a dull purplish to dark brown surface color. The sclerotium produces a nectar-like secretion known as "honeydew", and insects attracted to the honeydew aid in secondary passage of asexual conidia to other plants. Cool, moist weather and high concentrations of grasses or small grains in an area favor spread of the fungus and toxin production.

The life cycle of *Claviceps paspali* is similar to *C. purpurea* [5]. The sclerotia of *C. paspali* are irregularly rounded with a rough surface and are typically a tan to pale-yellow color. Dallis grass is grown primarily in the south central and southeastern USA and is not winter hardy for northern climates. Sclerotia may be consumed from pasture or when stored as hay. Paspalitrems are associated with warm humid climates of the southeastern USA, the natural range of Dallis grass. The ergot sclerotia of *Claviceps paspali* are visible by macroscopic examination of seed heads of infected Dallis grass [2, 3, 5]. Alkaloids of *C. paspali* include indole compounds that are derivatives of lysergic acid and known collectively as paspalitrems. Paspalitrems include paspalinine, paspalitrem A and paspalitrem B [2, 3, 5].

Forage Related Mycotoxins (Lolitrems, Ergopeptide Alkaloids, and Slaframine/Swainsonine)

Of three forage related mycotoxins important to horses, two are produced in a specialized relationship between endophytic fungi and important forage or pasture plants resulting in two distinct equine problems. Endophytes are fungi that live within vegetative tissue of plants and inhabit their seeds as well. The relationship between the plant and the fungus results in synthesis of known mycotoxins. These mycotoxins include ergopeptide alkaloids produced by the endophyte fungus *Neotyphodium coenophialum* growing in Tall Fescue (*Festuca arundinacea*) and *Neotyphodium lolii* growing in Perennial Ryegrass (*Lolium perenne*). The fungal infection is not visible macroscopically from the surface, but can be identified microscopically in plant tissue. An understanding of the conditions supporting their formation provides several means for prevention and control [5, 12].

Neotyphodium lolii in Lolium perenne (Ryegrass Staggers)

Neotyphodium lolii is an endophyte fungus of Perennial ryegrass (*Lolium perenne*) with production of four tremorgenic mycotoxins designated lolitrems A, B, C and D. The B fraction is the major toxin causing a condition called ryegrass staggers in ruminants and horses. The condition, known as ryegrass staggers, has long been recognized in Australia, New Zealand and Europe and has been previously found in the Pacific Northwest region of the USA [2, 5]. Toxin production is supported by hot dry weather, drought stress and overgrazed pastures, and toxin production appears correlated with higher levels of fungal infection. Toxin concentration is highest in the seeds and dispersal of infected seeds supports new infections in susceptible plants. Poisoning may appear several days after heavy dew or light rainfall. Toxin concentration is greatest in the lower 2 cm of the plant. Intensively grazed pastures during late summer or fall are more susceptible to infection from contaminated soil [37].

Rhizoctonia leguminicola in Trifolium pratense (Red Clover)

Red clover (*Trifolium pretense*) is frequently infected with a soil fungus (*Rhizoctonia leguminicola*) to cause a common plant disease known as "black patch". Less commonly other legumes (white clover, alsike clover, Korean lespedeza, cow pea, black medic, blue lupine) may also be infected. The infection appears as brown to black spots from 0.1 – 1 cm in diameter [5, 38-40]. High moisture and cool growing temperatures favor production of the toxins which include swainsonine and slaframine. A distinct cholinergic clinical syndrome of horses and ruminants known as Salivary Syndrome or "slobbers" is produced when the preformed slaframine is consumed from either fresh pasture or stored hay [38-40].

TOXICOSIS BY MYCOTOXINS: GENERAL PRINCIPLES

Dosage and Metabolism

Mycotoxicoses generally obey the toxicological principal of dose and response, and for each mycotoxin there is a threshold below which demonstrable effects do not occur in a given species [10-12]. Confusion sometimes arises for equine owners because effects may be acute or chronic and can be mild to moderate and subtle in their presentation. In fact, clinical experience suggests that mycotoxicoses often occur in a mild or chronic form. Response depends on the specific mycotoxin, the dosage and the time over which exposure occurs. Some mycotoxins are metabolized to products that are different in toxicity from the parent mycotoxin, and drugs or nutrients that alter foreign-compound metabolism may change the toxic response to mycotoxins [5, 6]. This is well documented for aflatoxins and ochratoxin, but appears relatively unimportant for trichothecenes, fumonisins and ergopeptide alkaloids.

Toxicity of Mycotoxins

Mycotoxins vary substantially in their toxicity on a body weight basis or by the concentration of mycotoxin measured in grain or forage. Toxicity values are sometimes expressed as mg/kg body weight (BW) or as a concentration (parts per million [ppm] or parts per billion [ppb]) in a dietary component. Table 2 compares 5 common mycotoxins that affect horses. Note that feed or forage concentrations are expressed primarily as ppm, but that the more toxic compounds (aflatoxins and ergovaline) are quite low and are commonly expressed in ppb. Where possible, a dietary concentration is also included. Conversions of body weight dosages to dietary concentration assume a daily intake of 20 % of body weight on a dry matter basis. One must also carefully evaluate a study where there may be a contaminated concentrate fed, but part of the daily ration includes hay or pasture. In general, the body weight dosage and concentration of a toxin in feed can be estimated by the formula:

Dosage (mg/kg BW) = Diet concentration in mg/kg diet X Percent body weight consumed

where diet concentration is often expressed as ppm and percent body weight is expressed as a ratio (e.g. 2% is expressed as 0.02). Remember that if a mycotoxin-contaminated grain is only half of the total diet, then the body weight dosage of mycotoxin is only half of what it would be if the entire diet were contaminated.

Table 2. Projected Response of Horses to Five Common Mycotoxins*

Toxin	Acute (<7 days)	Subacute (7 - 30 days)	Chronic (> 30 days)
Aflatoxin	2 mg/kg BW Diet = 100 ppm	0.075 mg/kg BW Diet = 3.75 ppm	Diet = 0.5 – 1.0 ppm
Deoxynivalenol	No acute studies	Grain = 11.0 ppm but only 41% of daily diet. Feed intake decreased	Barley in diet = 36-44 ppm; 1.27 kg consumed daily (approximately 10% of total diet). No effect on feed intake
Ergovaline	No acute studies	Fescue forage > 200 ppb (0.2 pppm) natural exposure	No reproductive effects if exposure stopped > 30 days prepartum
Fumonisin	0.01 – 0.05 mg/kg IV No dietary equivalent	0.6 – 4 mg/kg BW Diet = 30 – 200 ppm	Diet > 10 ppm
Slaframine	6 – 50 µg/kg BW by injection	NA	NA

*Compiled from information in references [1, 5, 16, 19, 22, 24, 29, 31, 40, 44, 45, 46, 57, 63, 66, 69, 70, 74, 76].

Example: The body weight dosage for a concentration of 0.5 ppm aflatoxin in contaminated grain would be mg/kg BW = 0.5 mg/kg diet X 0.02 = 0.010 mg/kg BW.

If only half of the daily diet(above) is grain, then the actual body weight dosage becomes lower as follows: mg/kg BW = 0.5 mg/kg diet X 0.01 = 0.005 mg/kg BW

Interaction of Mycotoxins

Two or more mycotoxins may potentiate the action of one another, or at least exert an additive effect if a similar mechanism of action or organ system is involved for both toxins [41, 42]. Currently there is limited evidence that common mycotoxins act synergistically, and there is virtually no evidence for or against mycotoxin interactions in horses. Generally, production of several mycotoxins at toxic concentrations at the same time in the same grain supply is uncommon [8, 9, 11]. Examples of co-occurrence are zearalenone and deoxynivalenol produced concurrently in corn or wheat, and fumonisins and aflatoxins detected together in corn.

Immunosuppression

Specific mycotoxins can alter immune function, but not all mycotoxins are associated with immunosuppression [5, 6, 10, 11, 13]. Aflatoxins, trichothecenes, fumonisins and ochratoxin A have been demonstrated immunosuppressive in domestic or laboratory animals. Of these, fumonisins are a relatively weak immunosuppressant, but aflatoxin has multiple effects on immune function. Trichothecenes may have effects that both stimulate or suppress specific immune functions [31]. Often immunosuppression by mycotoxins occurs primarily at concentrations that cause clinical toxicosis or when dosing or forced feeding experiments are conducted [11, 13, 31]. One should carefully evaluate research information that suggests broad and significant immunosuppression at subclinical exposure to mycotoxins. In addition, a serious challenge to accurate diagnosis occurs when a small amount of mycotoxin is found in a feed source concurrent with an infectious disease outbreak. There is sometimes a tendency to consider immunosuppression from a mycotoxin as the cause of these disorders. Because the immunosuppression could be reflected as increased prevalence or severity of infectious disease, mycotoxin-facilitated disease is difficult to detect or confirm. Field occurrence or research confirmation of mycotoxic immunosuppression in equines has been difficult to document under clinical conditions.

Clinical Toxicosis by Specific Mycotoxins

Clinical response of horses to mycotoxins may be acute, subacute, or chronic and both dose and time dependent, similar to other chemical toxins [1, 2]. An understanding of the range of effects for specific mycotoxins is important in differential diagnosis and evaluation of clinical prognosis for mycotoxin diseases. For known mycotoxins of clinical importance, response to some is subacute or chronic and the presenting signs are often subtle and vague (e.g. alterations of the reproductive cycle, reduced feed intake, slow growth, or impaired feed efficiency) [5, 6, 11, 13].

For the veterinary clinician, the challenge of mycotoxin diagnosis often begins with presentation of a vague acute or chronic condition that is not readily apparent as an infectious disease or physical condition. For the equine species there are relatively few well defined mycotoxicoses that occur with some frequency. Six mycotoxin groups important to equines worldwide include aflatoxins, fumonisins, ergopeptide alkaloids, slaframine-swainsonine, lolitrems and paspalitrems [4-7, 20]. Major effects of these mycotoxins include hepatotoxic damage and impaired growth (aflatoxins), acute liver disease and/or equine leucoencephalomalacia (fumonisins), agalactia, dystocia and prolonged gestation (ergopeptide alkaloids), mannose accumulation and parasympathomimetic signs (swainsonine and slaframine) and acute tremorgenic effects from specific grasses (lolitrems from *Acremonium lolii* and paspalitrems from *Claviceps paspali*) [2, 4, 20]. These will be described in some detail because of their current importance. Other mycotoxins with potential importance to horses, but not specifically identified as a clinical problem are characterized briefly. If poor quality feeds or forages appear associated with equine disease, these additional mycotoxins may be worthy of consideration provided their known effects match the clinical observations and other conditions are reasonably eliminated. For the less common mycotoxins, judicious use of laboratory results to confirm the toxicosis is quite important.

Clinical Characteristics of Specific Mycotoxicoses in Horses

Aflatoxins

Pathophysiology of Disease

Of the aflatoxins commonly recognized in grains (AFB_1, AFG_1, AFB_2, AFG_2), Aflatoxin B_1 is most toxic of the group followed in decreasing toxicity by AFG_1, AFB_2 and AFG_2. Aflatoxin absorption, metabolism and excretion are well characterized in many species, and although not specifically documented in horses, the known properties have relevance to equine clinical aflatoxicosis. Basic pathophysiological characteristics have been summarized. Absorption is by passive diffusion from the small intestine, with absorption rate greater in young animals. Plasma aflatoxin half life ranges from 12.9 to 36.5 in several laboratory animals, and it accumulates primarily in liver with additional amounts in kidney, bone marrow and several other tissues. The predilection for liver is important because Aflatoxin B_1 is metabolized primarily by liver and to some extent in kidney and small intestine. Hepatic microsomal mixed-function oxidases form seven major metabolites. Aflatoxin M_1, a major animal metabolite is excreted in both milk and urine. In acute exposure, blood and urine may contain both AFB_1 and AFM_1 residues. Toxicosis has been documented in piglets nursing sows exposed to aflatoxin. AFM_1 is rapidly excreted, and although specific equine data is scant, other domestic livestock excrete most residual aflatoxin within 4 – 7 days after exposure stops [4, 10, 19, 43].

The mechanism of action for aflatoxin is the result of an 8, 9-epoxide that binds covalently to DNA, RNA and proteins with reduced synthesis of DNA, nuclear and nucleolar RNA and protein. Aflatoxin is recognized as responsible for toxic signs and lesions, reduced growth, immunosuppression and in some species liver cancer [19]. Reduced protein synthesis and inability to mobilize fats produces characteristic lesions of hepatic fatty change and necrosis, as well as clinical evidence of liver dysfunction and changes in gastrointestinal tract and kidneys. Chronic low-level exposure may cause reduced growth rate. Protein-deficient diets exacerbate aflatoxin effects, and partial improvement in weight gain is provided by increased dietary protein [6, 7, 10, 43]. Aflatoxins may interact with other mycotoxins and either potentiate their action or at least act in an additive manner, for example in swine with the combination of aflatoxin and ochratoxins or aflatoxin and fumonisins [41, 42]. Specific evidence of aflatoxin interaction with other mycotoxins in horses has not been found, but these effects should not be discounted based on evidence from other species. Because both aflatoxins and fumonisins cause hepatocyte damage with similar clinical chemistry changes, the potential for an additive effect in horses exists if fumonisins and aflatoxins were fed simultaneously.

Clinical Signs and Toxicity

Relatively few natural cases of equine aflatoxicosis are reported, but experimental studies verify the risk to horses from aflatoxin consumption. Low reporting of aflatoxicosis in horses is not likely a result of greater resistance of equines to aflatoxin, since research studies suggest they are comparable to swine in susceptibility to toxicosis [16, 17, 44-46]. Selective feeding of higher quality grains or lower acceptance of damaged grains by horses could be a reason for the apparent lower risk to horses compared to other farm animals.

Liver is the major organ associated with acute or subacute aflatoxicosis, and signs in other organs (hemorrhages, immunosuppression and slow growth) may be in part related to the multiple roles of liver function in maintaining health. For acute exposures, intestinal tract and kidneys may also be involved [16, 47, 48].

Acute clinical disease is less common than subacute or chronic effects, but experimental acute toxicosis is a basis for the expected acute effects. Clinical signs can be similar to those for other diseases, but often include anorexia, depression, ataxia and dehydration, diarrhea with or without melena, straining, icterus, hemorrhage, and occasionally seizures [44-46, 49, 50]. Very high dosages (2 mg/kg body weight, equivalent to 50 ppm dietary aflatoxin) cause clinical signs within approximately 72 hours characterized by anorexia, fever, tachycardia, ataxia, colic, icterus, convulsions, bloody feces and abdominal straining [45]. As little as 0.075 mg/kg body weight for more than 35 days has caused death in ponies [46]. Assuming a feed intake at two percent of body weight, a dosage of 0.075 mg/kg is equivalent to a dietary concentration of 3.8 ppm – a very high dietary concentration of aflatoxins. Much lower concentrations of 500-1,000 ppb (0.5 – 1.0 ppm) are associated with reduced feed intake and loss of body weight [50]. Clinical signs at these concentrations are initially subtle with changes typical of advancing liver disease. These include reduced body weight, anorexia, depression, anemia, icterus and possibly petechial or ecchymotic hemorrhages in the subcutis or elsewhere [17, 44- 46]. Toxicity values reported in clinical cases have ranged from 55 ppb to 6500 ppb [3]. However, mycotoxin samples collected from field cases often do not reflect actual dosage received [30, 31]. Immunosuppression from aflatoxicosis has not been evaluated in horses, but other species with similar toxicity (such as swine) have experienced immunosuppression at feed concentrations generally between 200 and 400 ppb [11].

Clinical Laboratory Changes

Serum clinical chemistry results for acute experimental toxicosis have included prolonged prothrombin times (PT) and elevated serum concentrations of total bilirubin, ornithine carbamyl transferase (OCT), aspartate aminotransferase (AST), iditol dehydrogenase (IDT), gamma glutamyl transpeptidase (GGT) and arginase (ARG). Serum bile acids and icterus index may be increased, while bromsulphalein excretion is reduced [17, 44-46]. Animals that endure prolonged hepatic insufficiency may develop neurological signs consistent with hepatic encephalopathy.

Gross and Microscopic Lesions

Gross lesions documented in fatal experimental cases include icterus, hemorrhages, pale or fatty livers and hemorrhagic enteritis. Prolonged liver damage results in hepatic fibrosis hepatomegaly, ascites, hydrothorax and transmural edema of the gall bladder. Additional lesions associated with field cases have included pale swollen kidneys and hemorrhagic enteritis [16, 17, 44- 46, 49].

Microscopic lesions of aflatoxicosis are reported to include hepatocellular degeneration and necrosis, bile duct proliferation, fibrosis and hepatocellular megalocytosis with microscopic evidence of lipid infiltration of liver, myocardium and kidneys, and protein precipitate in renal tubules. In addition, gross and microscopic neurological lesions may include cerebral edema and neuronal degeneration [14, 16].

Diagnosis

Aflatoxicosis should be suspected when an acute or chronic disease involves apparent protein deficiency or dysfunction alone or with evidence of liver damage, especially if there is local history of aflatoxin contamination of crops or if there is recent change in health after purchase of new grain [6, 16, 17, 46]. A history of acute onset of anorexia with hepatic dysfunction and intestinal hemorrhage associated with damaged grain consumption could signal aflatoxicosis [1, 2]. Likewise, continuing ill thrift, weight loss, chronic liver insufficiency or unexplained immune suppression may suggest aflatoxicosis as a differential diagnosis.

Confirmation of diagnosis usually depends on characteristic clinical response, documentation of typical serum and liver changes, presence of high concentrations of aflatoxin in recent feedstuffs, and hopefully demonstration of aflatoxin metabolites in currently affected animals [12, 17, 46]. Because claims for damages often accompany evidence of contaminated feed, it is prudent to collect a thorough history and all possible animal and feed sources to document as well as possible the claimed damages. Thorough documentation of the circumstances with identification and preservation of samples is a valuable service by the attending veterinarian. Hematology and serum biochemical changes will confirm acute or chronic liver damage. Analysis for aflatoxins in grain or feed will aid in establishing exposure, but only if contamination is recent [11]. Horses have not been evaluated for residues of aflatoxin metabolites, but from data in animals with similar susceptibility, aflatoxin B_1 persists for only a few days in blood or urine and the metabolite Aflatoxin M_1 may be retained for less than three weeks. Thus, diagnosis from aflatoxin residues in urine or tissue must be done promptly after exposure stops. At lower concentrations in feed, the slow onset of signs may be apparent only after the offending source is gone [4, 10].

Treatment

Specific antidotes are not available for aflatoxicosis. Acute aflatoxicosis has been partially mitigated with a combination of L-methionine (200 mg/kg body weight) and sodium thiosulfate (50 mg/kg) started 8 hours post exposure. If hemorrhages are present, Vitamin K_1 should be included in the treatment regimen [3]. However, these approaches are difficult to accomplish on a herd basis and have not been well tested in horses.

Management of surviving animals generally includes removal of all known feeds, supplementation with fat soluble and B vitamins, selenium supplementation, and ample provision of high quality easily digested protein. If consumption of contaminated grain is unavoidable, contaminated diets can be treated with 5 to 10 lbs per ton of hydrated sodium calcium aluminosilicate (Novasil®) which effectively adsorbs much of the aflatoxin. In food animals and dairy cattle this has improved performance and reduced residues by as much as

50 – 80%. Some bentonites have produced good results in contaminated diets of dairy cattle and sheep, but aluminosilicate adsorbents appear to have been tested in horses [4, 12, 43].

Ammoniation of commodities such as corn and cottonseed hydrolyzes the lactone ring of Aflatoxin B_1 to less toxic fractions [51]. However, this process is not documented for horses and does not have full FDA-CVM approval. Organic acids added to grain will prevent mold growth, but do not destroy aflatoxin.

Recovery and Prevention

Recovery from aflatoxicosis can involve long term nursing and should include excellent nutrition with high quality protein diets and supplemental vitamins. Post exposure liver evaluation may help to determine if hepatic recovery is likely.

Prevention of aflatoxicosis requires attention to grain quality and feed management. Be aware of regional prevalence of aflatoxin and consider laboratory verification of freedom from aflatoxins when purchasing grain for equine diets. Store grains and feeds in clean, dry facilities at approximately 12% grain moisture to prevent aflatoxin production in storage. Representative sampling of large grain supplies is an important factor for accurate evaluation of suspect aflatoxin contamination [10-12]. Hydrated sodium calcium aluminosilicate (HSCAS) added to feeds has reduced the effects of aflatoxins in swine, cattle and poultry [6, 11]. While comparable information is not available for horses, the aluminosilicate adsorbents would be a logical alternative if a contaminated grain must be fed. U S Food and Drug commodity advisories do not specifically address aflatoxin limits for horses but based on other species a concentration less than 100 ppb dietary aflatoxin for adult horses and less than 50 ppb for foals is a rational limit [43].

Fumonisins

Toxicity and Mechanism of Action

The role of fumonisins in sphingolipid synthesis is currently demonstrated to be a factor in their toxicity and carcinogenicity [21, 29, 52-55]. Fumonisins inhibit the synthesis of sphingosine by inhibiting ceramide synthase, the enzyme which converts sphinganine to sphingosine. As a result, the ratio of sphanganine to sphingosine rises several-fold and may be useful in early detection of fumonisin exposure. Elevated sphinganine/sphingosine ratios are associated with fumonisin toxicosis. Alteration of sphingolipid metabolism is proposed as a factor in the cause of ELEM and/or hepatotoxicosis from fumonisins [21, 29, 54, 55]. *In vitro* and laboratory animal studies suggest that fumonisins may reduce the barrier function at junctions of endothelial cells, potentially leading to leakage of fluids and proteins into the central nervous system [54, 56]. In addition, although liver damage is a common feature of fumonisin poisoning, no mechanism as yet accounts fully for the hepatotoxicosis that often precedes the neurological signs of equine leukoencephalomalacia (ELEM).

Horses are highly susceptible to Fumonisin B_1 (FB_1). Feed concentrations above approximately 8 ppm for more than 30 days are believed capable of causing ELEM [57]. For higher dietary concentrations, clinical signs can develop within 1 – 2 weeks after exposure begins, while low or intermittent concentrations are believed to require several months to cause clinical signs. An early biomarker of fumonisin effects is an elevated ratio of serum sphinganine to sphingosine because of the inhibition of this conversion. Cellular membrane function is impaired and believed responsible for edema and protein passing the blood brain barrier [29, 55]. Horses are most sensitive among the domestic animals and dietary concentrations as low as 8-10 μg/g (ppm) for more than 30 days have been associated with fatal ELEM [57]. Onset of clinical signs appears to be dose dependent and can range from 7-10 days at high fumonisin concentrations to several months at low dietary exposures [22, 57, 58]. High dosages of fumonisins appear to cause predominantly liver toxicosis, while lower exposure for longer time periods favors ELEM [22, 29].

Clinical Signs

ELEM is often preceded by mild to moderate liver dysfunction which may not be evident clinically. Although CNS signs may be delayed for weeks during exposure, once they appear there may rapid progression to severe effects that are almost always fatal. There is ataxia, wandering, circling and head pressing. In some cases signs of stimulation that include excitement and muscle tremors may predominate. Impending terminal signs include depression, severe ataxia, stumbling, paresis, facial paralysis, weakness or protrusion of the tongue, blindness, coma, and death. Once affected, horses rarely survive longer than 5 – 7 days [6, 20, 57, 59]. Icterus from liver disease may accompany the CNS signs. Affected animals rarely survive, but any that do would be expected to have permanent brain damage.

Only equidae are known to develop the brain lesion of equine leukoencephalomalacia (ELEM) under natural exposures. Clinical effects of ELEM include rapidly progressing neurotoxicosis characterized by depression, blindness, ataxia, aimLess wandering, head pressing, circling, hyperexcitability, facial paralysis, paresis, coma and death. Affected animals may stand motionless for long periods of time, or show continuous repetitive movements until death. Tonic-clonic seizures may occur in recumbent animals. Sudden death may be the first sign observed [3, 6, 20-22, 27, 59]. Time from onset of clinical signs to death may be as little as 24 hours to as long as a week. Concurrent hepatotoxicosis with elevated serum liver enzymes and icterus may occur [22, 57, 58]. ELEM is fatal in nearly all affected animals, and any clinically affected surviving animals are likely to have substantial neurological impairment.

High to moderate dosages of fumonisins cause liver failure with signs that include icterus, hemorrhages, edema and depression. Serum biochemical changes include elevations in alkaline phosphatase, aspartate amino transferase, lactate dehydrogenase, gamma glutamyl transpeptidase, and bilirubin. Liver enzymes typically rise in a dose related manner after 7-10 days exposure to fumonisins [22, 24, 57, 58]. In addition, severe hepatotoxicosis is associated with increases in serum ammonia and possible hepatic encephalopathy [20].

Lesions

Fumonisin hepatotoxicosis appears as swollen and brownish discolored livers with irregular foci or nodules throughout the liver [57, 58, 60]. Microscopic examination reveals prominent hepatocyte vacuolation and fatty change in hepatocytes, loss of hepatic architecture, hepatocellular apoptosis, periacinar necrosis, portal fibrosis, bile stasis, and bile ductule proliferation. Individual and scattered liver cell necrosis may occur randomLy throughout the liver. Liver lesions may be reversible if exposure to fumonisins ceases. If prolonged liver damage and fumonsin exposure continue, ELEM may also develop [29, 55].

ELEM appears grossly as massive softening and liquefaction of subcortical cerebral white matter (Figure 2). Lesions may range from discrete focal areas to large cavitations with inward collapse of the cortical gray matter. Lesions range from very small (1mm) to 4 cm or more, sometimes involving nearly the entire white matter of either or both sides of the cerebral hemispheres. Hemorrhages are prominent, and microscopically there may be liquefaction of white matter. Lesions are usually not bilateral or symmetrical and may also occur in the thalamus, cerebellum, brain stem or medulla oblongata [3, 20, 57, 58].

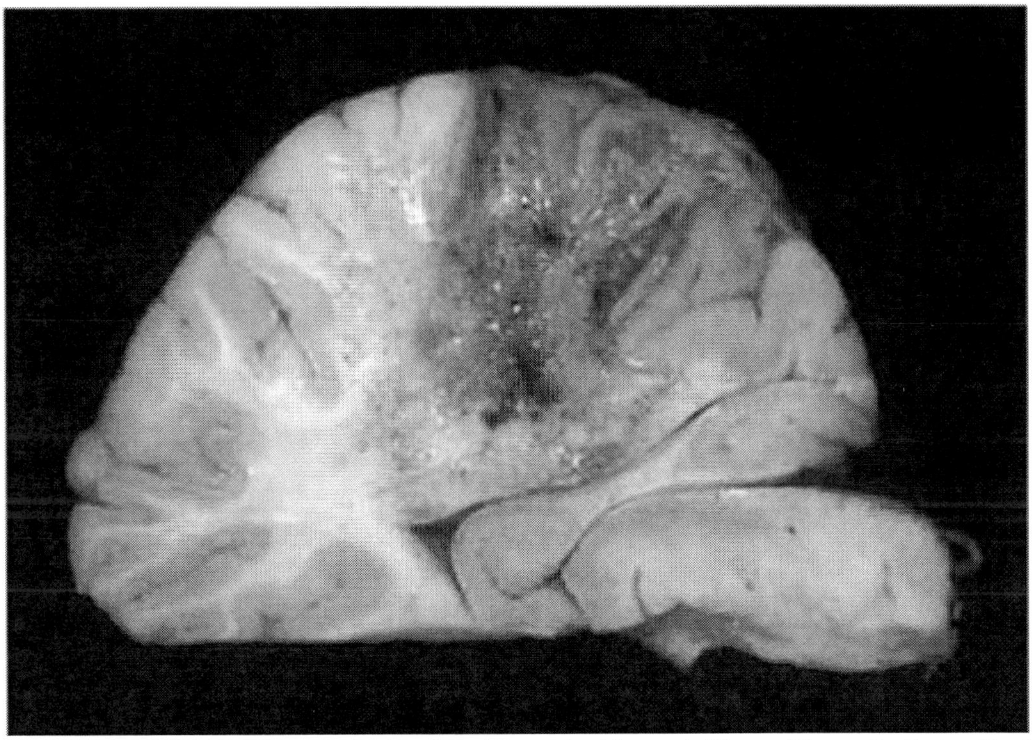

Photo courtesy of C.B. Hong, The University of Kentucky Livestock Disease Diagnostic Center. Used with permission.

Figure 2. Transverse section of equine brain with lestion of equine leukoencephalomalacia (ELEM). The lesions include multiple focal hemorrhages of the cerebral white matter and midbrain; as well as corgulative to liquafactive necrosis and of cerebral white matter with extension through the cortical gray matter.

Microscopic brain lesions are characterized by loss of cerebral architecture and cellular morphology [2, 3, 20, 29, 57]. The necrosis is nearly always confined to white matter with liquifaction necrosis and accumulation of macrophages. Glial cells may have pyknotic nuclei and eosinophilic cytoplasm. Perivascular cuffing and proliferation of macrophages may occur in response to the necrosis. Cerebrospinal fluid may be yellow-tinged with increased total protein and elevated inflammatory cells and erythrocytes.

Clinical Laboratory Changes

Clinical chemistry changes typical of fumonisin toxicosis reflect liver damage. High values are reported for AST, LDH, GGT, cholesterol and bilirubin. For prolonged ELEM, the more acute phase of liver damage may have passed by the time death occurs [2, 3, 20, 29, 57].

Diagnosis and Laboratory Evaluation

Fumonisin toxicosis should be considered if hepatic dysfunction or acute neurological signs appear in horses or other equidae. History of feeding or foraging on damaged or molded corn should be determined, although the long latent period before clinical signs makes accurate history difficult to obtain.

Early changes in fumonisin-poisoned horses often include hepatic dysfunction. Serum enzymes and constituents affected by fumonisins include increases in serum lactate dehydrogenase, alkaline phosphatase, aspartate aminotransferase, gamma glutamyl transpeptidase and bilirubin [20, 29, 57, 58]. Serum sphinganine/sphingosine ratio is increased several fold and may be the earliest and most sensitive indicator of fumonisin exposure [52, 55]. Sphingolipid assay is relatively difficult, and few clinical laboratories or veterinary diagnostic laboratories offer this analysis. Direct analysis for fumonisins in tissues or urine is not usually done, as fumonisins appear to be excreted readily and rapidly in both bile and urine of species studied, although specific kinetic studies in horses were not found [55].

Diagnosis

Typical neurological signs with a history of corn consumption should suggest ELEM as a differential. Serum clinical chemistry indicating liver dysfunction would increase the probability of ELEM. For deceased animals, the CNS lesions are nearly pathognomonic. Analysis for fumonisins in available feed should be included in the diagnostic workup. However, specific concentrations of fumonisins in corn collected after a poisoning may not accurately reflect total fumonisin dosage because of the often extended time period for exposure and the delayed nature of the poisoning. If available, testing serum for increased sphinganine/sphingosine ratio is highly suggestive of fumonisin exposure [29, 57, 59].

Treatment and Prevention

There is no antidote for fumonisin toxicosis and treatment of ELEM is essentially futile. Because of multiple days of exposure for toxicosis to occur, oral decontamination is usually of limited value. Clinically affected horses with ELEM offer a grave prognosis. Control of hyperactivity or ataxia to prevent trauma is probably most important. If clinical diagnosis is accurate and a realistic prognosis is given, euthanasia may be the owner's choice. Animals with liver damage should be provided appropriate supportive care for identified hepatic damage [2, 20].

For prevention, corn can be analyzed for fumonisins to confirm potential toxic concentrations. Contaminated corn can be cleaned and the good-quality grain analyzed to confirm the removal of toxic residues [24, 30]. Preventive measures include minimizing the amount of corn in equine diets, using only cleaned corn that removes contaminated screenings, or testing each lot of corn used for fumonisins.

Regulatory controls are currently not formally enforced, but advisory levels are established, and FDA currently recommends no more than 5 ppm dietary fumonisins in equine feeds, which should constitute no more than 20% of the total diet [61]. Fumonisins are potential human and animal carcinogens, and their eventual regulation in foodstuffs seems likely.

Comprehensive Control and Prevention of Fusarium Mycotoxin Problems

A comprehensive long- term and industry wide HACCP approach has been described by Jouany [62] that provides a substantial and thoughtful approach to preventing and controlling *Fusarium* fungal and mycotoxin contamination in crops, and the interested reader is encouraged to consult this readily available online reference.

Deoxynivalenol and Other Trichothecene Mycotoxins

Group A Trichothecenes

Experimentally the group A trichothecenes (T-2 and DAS) cause dermal irritation and necrosis, lymphoid depletion, gastroenteritis, diarrhea, shock, cardiovascular failure, and death. Chronic administration results in reduced hematopoiesis and pancytopenia. A and B group trichothecenes are reported immunosuppressive [6]. Because T-2 and DAS also cause marked feed refusal (data from swine and rat studies) their toxic effects are considered somewhat self limiting except for reduced feed intake [63]. While much experimental work has been done with the Group A trichothecenes (T-2 and DAS) in swine and laboratory animals, they occur only occasionally and rarely at toxic concentrations in North American grains. Recent reports of naturally occurring Group A trichothecene toxicosis in horses are virtually unknown [6, 31].

Attempts to prevent Trichothecene effects have centered on feeding adsorbents, or chemical or physical means of detoxification [30]. These have included calcium aluminosilicate, bentonite and sodium bisulfite. These have generally been unsuccessful as an economical or practical prevention in any species, and no reports were found for horses.

Group B Trichothecenes

Recent experimental evidence has provided some insight into the effect of Group B Trichothecenes (DON and 15-acetyl DON) on horses and the response of horses to glucomannan adsorbent agents (GMA) against *Fusarium* mycotoxins. Information relevant to horses is summarized here for this recent clinical perspective.

Johnson et al. [64] provided five healthy adult horses DON-contaminated barley for 40 days at the rate of 1.364 kg twice daily. Horses were kept in separate paddocks and allowed free choice water and grass. The barley was analyzed and confirmed free of other specific mycotoxins (aflatoxin, T-2 toxin, diacetoxyscirpenol, ochratoxin A, citrinin, and zearalenone). Concentration of DON in the barley ranged from 36-44 μg/g (ppm) throughout the trial. All horses were observed for clinical signs, and feed intake was recorded daily.

Blood was collected for hematology and serum biochemical assays at ten-day intervals and serum IgA and IgG were measured by single radial immunodiffusion. All horses remained clinically normal during and after the trial. There were no differences in feed consumption throughout the study and no evidence of feed refusal in the five horses fed contaminated barley. The authors indicated "no remarkable changes" in hematological or serum biochemistry parameters. Although hematocrit and serum enzyme activities decreased slightly, all values were within normal reference ranges, and there were no biochemical indications of hepatic damage. Unlike some other experimental animals, horses did not experience increased serum IgA with exposure to DON. The authors' original hypothesis included that DON might promote glomerulonephritis, as has been reported for experimental studies in mice [65]. Based on no observed adverse clinical effects, normal hematology and serum biochemical values and failure to increase IgA in experimental animals, the authors concluded that horses may be less susceptible than other monogastrics to the effects of DON. By comparison, swine experience reduced feed intake at 1 ppm dietary DON and may demonstrate complete feed refusal at DON concentrations between 5 and 10 ppm [11].

Raymond et al. 2003 [66] fed horses a blend of grains naturally contaminated with *Fusarium* mycotoxins (average 15.0 ppm deoxynivalenol, 0.8 ppm 15-acetyldeoxynivalenol, 9.7 ppm fusaric acid, and 2.0 ppm zearalenone) to evaluate effects on feed intake, serum immunoglobulin (Ig) concentrations, serum chemistry, and hematology of horses and to determine the effect of adding a polymeric glucomannan mycotoxin adsorbent (GMA). Nine mature, mixed-breed mares were assigned randomLy to one of three dietary treatments for 21 d and the experiment replicated in time following a 14-d washout interval. Each mare was provided 2.8 kg of concentrates and 5 kg of timothy/alfalfa hay. The concentrates were 1) control, 2) blend of contaminated grains and 3) blend of contaminated grains + 0.2% GMA. Diets containing contaminated grains average 15.0 ppm of deoxynivalenol, 0.8 ppm of 15-acetyldeoxynivalenol, 9.7 ppm of fusaric acid, and 2.0 ppm of zearalenone. Feed intake by all horses fed contaminated grains was reduced ($P < 0.001$) compared with controls throughout the experiment. Supplementation of 0.2% GMA to the contaminated diet increased ($P = 0.004$) feed intake of horses compared with those fed the unsupplemented contaminated diet. Body weights of horses were unaffected by diet. Serum activities of γ-glutamyltransferase were higher ($P = 0.047$) in the contaminated diet compared to controls (P= 0.027) on day 7 and 14, but not on day 21 ($P = 0.273$). Supplementation of GMA to the contaminated diet decreased ($P < 0.05$) serum γ-glutamyltransferase activities of horses compared with those fed unsupplemented contaminated diet on 7 and 14 days. Other hematology, serum chemistry and immunoglobulin measurements were not affected by diet. The authors suggested that the increase in the activity of this enzyme indicates either hepatocellular damage or enzyme induction. Since the differences disappeared by 21 days it was suggested that the horses may have adapted to the hepatotoxicity caused by the combination of *Fusarium* mycotoxins. Another possibility could be that GGT activity is reduced by the GMA, which could be assessed by a GMA plus control grain treatment group.

Raymond et al. 2005 [67] fed blends of grains naturally contaminated with *Fusarium* mycotoxins to mature, exercised horses, and to test the efficacy of a polymeric glucomannan mycotoxin adsorbent (GM polymer) in preventing *Fusarium* mycotoxicoses. The contaminated grains averaged 11.0 ppm deoxynivalenol, 0.7 ppm 15-acetyldeoxynivalenol, and 0.8 ppm zearalenone on an as-fed basis. Three dietary treatments were 1) Control (no mycotoxins); 2) mycotoxin-containing concentrates, and 3) mycotoxin containing grains amended with 0.2% glucomannan mycotoxin adsorbent (GMA). Horses on a fixed exercise schedule throughout the experiment were fed 3.5 kg of concentrates and 5.0 kg timothy/alfalfa hay with feed intake and body weight monitored for a 21 day period. Exercise variables tested included 1) time-to-fatigue, 2) heart rate, 3) hematology, and 4) serum lactate concentration. Feed intake decreased in horses fed contaminated grain with or without 0.2% GM polymer compared with those fed the control diet throughout the study. Feed intake of horses fed contaminated grains had significantly ($P < 0.05$) decreased feed intake compared to controls, but the group with GMA added was not different in feed intake compared to the unsupplemented contaminated diet. However, the contaminated group received 11.5 ppm DON and the GMA plus mycotoxins group received 14.5 ppm DON. There were no effects of diets on hematology or variables of athletic ability. The authors reasoned that failure to show a differential effect from GMA might have been due to the higher concentration of DON in the GMA supplemented group.

Glucomannan adsorbents were tested against a naturally occurring combination of DON, 15-Acetyl DON, Fusaric acid and zearalenone in dogs. There was no protective effect from glucomannans against any of the four Fusarium mycotoxins evaluated [68].

These recent DON studies represent different species, and results are suggestive but not entirely consistent. They illustrate the need and opportunity to further evaluate different species and combinations of adsorbents in a search for some common or clarifying understanding of issues with trichothecene effects on feed consumption and other physiological or immunological functions.

Paspalum Tremorgens

Claviceps paspali is the source of several related mycotoxins collectively known as paspalitrems. Their probable mechanism of toxicosis is the inhibition of gamma amino butyric acid (GABA) or glutamate, two common inhibitory neurotransmitters [2, 3, 5]. Their absence allows presynaptic motor release and depolarization with facilitated transmission and predominance of excitatory effects in the central nervous system. Clinical effects of paspalitrems appear within three to seven days of grazing on infected pastures and are nearly identical to those of lolitrem B (See Ryegrass Tremorgens) although effects and clinical course are considered less severe than for ryegrass tremorgens [2, 3].

There is no antidote for pasalum tremorgens. Remission of signs should occur within 1-3 weeks after removal of animals from toxic pastures. Mowing of pastures will reduce access to sclerotia in the seeds and reduce the probability of toxicosis.

Ryegrass Tremorgens

Toxicity and Clinical Effects

Lolitrem toxins are characterized by an indole moiety as part of a complex ring structure. Various derivatives of lysergic acid, including the tremorgenic mycotoxins, are associated with reduced effects of the inhibitory amino acids such as gamma amino butyric acid

(GABA) and glycine. Loss of inhibitory amino acids leads to increased presynaptic neurotransmitter release and prolonged depolarization. Synaptic transmission at the motor end plate is facilitated providing a probable mechanism for the uncontrolled tremors and incoordination observed clinically [2, 3, 5]. Toxicity is not well defined for domestic animals, but plant concentrations of lolitrem B exceeding 2 μg/g (ppm) of dry matter are considered to cause signs [69].

Dramatic acute neurological effects characterize clinical signs from these tremorgenic mycotoxins. Effects are related to altered neurotransmission in the central and peripheral nervous system. Signs are typically mild and subtle when animals are at rest, but excitement or forced exercise causes exaggerated and severe response. Early effects include fine tremors about the head and neck. Stiffness, ataxia, hypermetria, tremors, opisthotonus and seizures develop within a few minutes, but may subside quickly if animals are left undisturbed. During tremorgenic episodes, animals may become belligerent and can be a threat to persons nearby. Severe neurologic signs can result in self-induced injury by running through fences, falling over embankments or drowning in shallow water. Horses may be affected by consuming either toxic pastures or hay, although most cases are reported on toxic pastures. Morbidity is typically high (> 50%) but mortality in uncomplicated cases is low (< 10%). Animals are expected to recover within 3-7 days if removed from contaminated forages and provided nursing care, food and water. Neither neurological nor muscle lesions are produced by tremorgenic mycotoxins. Elevations of serum aspartate aminotransferase and hemorrhagic lesions in musculature may be secondary to incoordination, staggering, falling and related muscle trauma [2, 3, 5].

Diagnosis is usually based on observation of typical clinical signs, history of exposure, and elimination of other differential diagnoses. Analysis for the tremorgenic mycotoxins is not commonly offered in diagnostic laboratories, and analysis is only likely in suspect forage or hay because the tremorgens are retained only a short time in tissues and at very low levels. *Acremonium lolii* may be detected by microscopic examination of infected ryegrass. A mouse bioassay can be used to aid detection of toxin in plant material. Suspect materials are ground and suspended in saline, then administered orally at approximately 0.5 mL/mouse. Signs of tremors and seizures appear within 10 to 30 minutes [2, 3, 5].

There is no antidote for perennial ryegrass staggers. Activated charcoal and saline cathartics for oral decontamination may be helpful, but are difficult to administer on a herd basis or in excitable animals. In most cases, removal from the suspect forage and nursing care in a quiet secure place will allow recovery within 3-7 days. Affected or recumbent animals usually will eat and drink, and nursing care should include protection from the sun and provision of fluids to prevent dehydration during recovery. Medications which increase CNS concentration of glycine (e.g. mephenesin; nalorphine) are reported to have reduced tremorgenic mycotoxin effects [2, 3].

FESCUE ENDOPHYTE TOXICOSIS

Fescue endophyte toxicosis in equines is a serious medical and reproductive problem when it occurs. It is generally limited to the regions where fescue is grown and used for horses or where hay is purchased from an endemic area. Horse owners and stables should be aware of the type of hay purchased, especially if it is transported from other areas. This overview is provided to call attention to major points in susceptibility, mechanism of action, toxicity, clinical effects, diagnosis and treatment. Excellent and detailed reviews are available for a more comprehensive treatment of this topic [70, 71, 72, 73].

Toxicity and Mechanism of Action

The principle toxic effects from fescue endophytes in horses are due to the ergopeptide alkaloids. These include ergovaline, ergosine, ergotamine, ergocristine, ergocryptine and ergocornine. Of these, ergovaline accounts for a reported 90% of the ergopeptide alkaloids present [72]. The toxicity arising from contaminated fescue can also be caused by ergotized grains or grass seed heads of common pasture grasses [74]. Concentrations of ergovaline in fescue forage may range from 200 – 600 μg/kg (ppb) and to as much as 1000 ppb in seeds [71]. Signs of agalactia in mares have been reported from forage concentrations of ergovaline at 100 ppb or more, and effects on lactation and pregnancy in the mare are the principal effects. Toxicity of tall fescue itself can change with year, season, drought stress and nitrogen fertilization [70, 71]. Thus, risk of toxicosis is dependent on weather and cultural factors that influence the ergovaline content of the forage. In addition, the period of greatest susceptibility in mares is during the last 30 days of pregnancy, and removal of mares from fescue before that period greatly reduced the risk of foaling abnormalities.

Ergovaline is a strong prolactin inhibitor and important in maintaining a functioning corpus luteum and gonadotropin secretion [71, 72]. Fescue forages are associated with low progesterone and high estradiol levels in mares [75]. These imbalances are associated with late gestation problems in mares. Stimulation of D_2 dopamine receptors by ergovaline decreases prolactin secretion, which leads to changes in utero-fetoplacental steroid metabolism or by yet uncharacterized changes in the fetal anterior pituitary [76]. Enlarged and edematous placentas in mares consuming toxic fescue have been associated with possible vasoconstriction, and changes of edema, fibrosis and mucoid degeneration present in the placenta [77].

Clinical Signs

Clinical effects of ergopeptide toxicosis in mares are evident near or at the end of gestation or in the early post partum period. The gestation period is typically extended 20-27 days and there is lack of the usual signs of impending foaling (udder development, dropping of the paralumbar fossa, relaxation of the tail-head and vulva, waxing of colostrum) [71, 72, 75, 77]. This means many foalings will be unexpected and without the usual preparation. Abnormal presentation of the foal, lack of preparation of the mare and grossly larger foals all contribute to Dystocia that threatens survival of the foal and possibly the health of the mare. Placental thickening occurs within the last 6.5 hours before parturition [78].

Reproductive dysfunction is summarized succinctly by Wright [73]. Specific changes include:

- extended gestation period from 11 to 12 months,
- dystocia, with mares sometimes trying to foal for many hours,
- agalactia with poor quality colostrum (low immunoglobulin levels),
- "red bag" placentas from premature separation, with the chorioallantois delivered before foal
- thick edematous placentas with weights exceeding 6.5 kg for a thoroughbred mare,
- weak or dead foals with aspiration pneumonia from struggling to get out through a thickened placenta.

Fetal asphyxia, weakness, dysmaturity (oversized foals) and mortality in foals also is reported [71, 75]. Once delivered, the surviving foal will be without colostrum or even a source of nutrient because of the profound agalactia of the mare. Unless this is corrected promptly the foal can have reduced immune competence or die from starvation. Foals born without the normal increases in maternal progestagens suffer hypoadrenocortical function and are small, weak or stillborn [79]. Edema of the placenta increases the placental weight.

Diagnosis

Laboratory tests for ergovaline (the ergot alkaloid of fescue toxicity) and lolitrem B (the ergot alkaloid of perennial ryegrass) are available in selected veterinary diagnostic or research laboratories. Some laboratories also offer assays for other ergopeptide alkaloids. Since ergopeptide alkaloid sources may include ergotized grains (and ergotized grass seed heads) the ability to detect a range of ergopeptide alkaloids is useful for differential diagnosis between fescue toxicosis and ergotism. At present, some tests for detection of ergopeptide alkaloids in urine are under development using HPLC, TLC or ELISA technology, but the alkaloids pass quickly from blood to urine and are often undetectable even in urine within two days after exposure stops [71, 72, 75]. Thus, testing is most effective if it occurs during an ongoing clinical toxicosis or active exposure.

Treatment

An early treatment or intervention is removal of late gestation mares from the toxic fescue. If removed 30 days before parturition, agalactia and foaling problems are almost always avoided. If gestation is prolonged but the mare is removed from fescue, prolactin will begin to increase within 24 hours and mammary gland development is detectable by the second day after removal [71, 72, 75]. However, up to a week may be required for redevelopment of the udder and full lactation. Mares can be treated with Dopamine D_2 antagonists (domperidone) for mares that are past their foaling date. Domperidone is recommended at 1.1 mg/kg orally SID for one week before parturition. It is marketed as an 11% gel (Equidone® from Equi-Tox Central, SC). If the mare has already foaled, the same dosage is given twice daily until milk flow commences [71].

Prevention

If mares must be kept on infected fescue pastures during late gestation, domperidone can be given at the treatment dosage during the last 10-14 days before expected parturition [78]. As indicated earlier, removal from fescue 30 days prior to expected parturition date will usually prevent the dystocia and agalactia problems. Wright [73] has recommended several specific measures for prevention of equine fescue toxicosis. Late-gestation mares should never be allowed to consume cereal-rye straw, endophyte-infected hay or stubble or eat sclerotia-containing grain. Tall fescue and perennial ryegrass sources for mares must be guaranteed endophyte-free. If endophyte free fescue is not available, remove mares from the pasture no later than 300 days of gestation to avoid serious foaling or agalactia problems. Mow grass pastures to keep them in a vegetative state rather than letting them develop seed heads, which reduces the development of airborne ascospores and reduces spread of the infection. Finally, arrange with a laboratory for preliminary testing for ergovaline in adequately selected samples pasture or hay. HPLC methodology is considered more sensitive than thin layer chromatography (TLC) when trying to identify ergot alkaloid concentrations less than 1 ppm.

Recommendations for sampling forages in contaminated pastures have been described [75]. For endophyte detection, approximately 30 sample sites per field or 5 sample sites per acre are suggested for forages or seeds. Collection time should be between May and December. Three stems per sample are taken and placed in a test tube half filled with water. Staining procedures or ELISA can be performed by a qualified laboratory. In addition, direct chemical analysis for ergovaline is available at selected veterinary diagnostic laboratories.

SLAFRAMINE

Toxicity and Clinical Effects

The observed effects from infected red clover consumption by horses appear related to slaframine. Slaframine is most often associated with second cutting red clover hay. Hay storage may result in degradation of slaframine over time, with values 10% of initial after 10 months of storage [39, 40].

Slaframine has strong parasympathomimetic effects. It is bioactivated in liver to a ketoimine with properties similar to acetylcholine, which stimulates muscarinic receptors [38, 39]. Biologic response of several species (horses, cattle, sheep, guinea pig, and swine) is similar. Parasympathetic stimulation is characterized most prominently by excessive salivation. Lacrimation, frequent or excessive urination and diarrhea are frequently reported [39, 40, 80]. Bradycardia and bradypnea occur as well. Excessive salivation may begin 10-30 minutes after ingestion of contaminated hay or pasture and may continue for one to three days. Observation reveals that several quarts to several gallons of saliva may be produced within a few hours as excessive salivation continues (Figure 3).

Courtesy of *The Draft Horse Journal* and Dr. A. J. Neumann. Used with permission.

Figure 3. Example of excessive salivation in a horse caused by slaframine produced in molded clover pasture.

Atropine given after signs occur does not alleviate the clinical signs. Animals may be anorectic for several days and experience mild colic in association with diarrhea and other parasympathetic signs. Moderate weight loss from anorexia and dehydration may be expected, but mortality is very rare and complete recovery would be expected when the affected pasture or hay is removed [2, 4, 5, 10]. Slaframine toxicosis is dramatic but rarely serious if exposure is stopped. Prevention is accomplished by recognizing the black patch disease and preventing exposure to susceptible animals.

Prevention of Mycotoxin Poisoning

Specific treatments or antidotes are not available for mycotoxins. Supplementation with fat-soluble and B complex vitamins and selenium is often recommended. High quality dietary protein and supportive nutrition for liver damage is recommended for aflatoxin and fumonisin hepatosis. Calcium aluminosilicates are useful for reducing effects of aflatoxins in swine, poultry and cattle but of limited or no value for other mycotoxins. Studies with aluminosilicate adsorbents in horses have not been found, but one could assume some level of efficacy against aflatoxins based on comparative results in three other species. Adsorbent agents for trichothecenes, zearalenone and fumonisins are of limited practical value [30] but recent trials with of glucomannan adsorbents in horses offer some promise for relief from the combined effects of DON, 15-Acetyl DON), fusaric acid and zearalenone [66, 67].

Dilution of contaminated grain will reduce the initial mycotoxin concentration, but molded grains may contaminate sound grain if storage conditions promote mold growth. Cleaning of grain to remove damaged and broken seeds will substantially reduce mycotoxin concentration because individual damaged seeds are often heavily contaminated. Heavily molded grains may lack normal energy and vitamin levels. Thus, careful attention to nutrient analysis with appropriate supplementation is advised [11, 12].

Storage of grain and feed should be at 12-13% moisture to prevent fungal growth. Organic acids will prevent mold growth, but do not destroy preformed toxins. Storage bins and feeders should be kept dry and cleaned periodically with dilute sodium hypochlorite. Milled feeds are susceptible to mold invasion, and should not be stored under conditions of high humidity and temperatures. Periodic testing of feeds and grains may be important if storage conditions have been questionable [11, 12, 30].

Conclusion

Horses consume feed grains and forages that can produce a range of mycotoxins resulting from mold invasion. Toxicosis of horses is most likely to occur from fumonisins or aflatoxin in grains and from the tremorgenic mycotoxins in dallis grass or slaframine in red clover. Fumonisin toxicosis is often severe and fatal and aflatoxin can be acute or chronic and debilitating. Fescue endophytes produce severe losses in pregnant mares and their foals and aflatoxins can cause effects from low productivity to severe or life-threatening hepatic failure. Other mycotoxins reported in horses may cause moderate to mild signs that regress when the contaminated feedstuff is removed. Overall, horses appear to have a relatively low prevalence of reported mycotoxicoses among the domestic animals, but they are exquisitely sensitive to the fumonisins and less so to ergopeptide endophytes and aflatoxins. With few good therapies for mycotoxin poisoning, attention to providing high quality grains and forages to prevent mycotoxicoses and early recognition of impending problems are the most effect means for reducing risk from mycotoxins in horses.

References

[1] Hintz HF. Molds, mycotoxins, and mycotoxicosis. *Veterinary Clinics of North America: Equine Practice*, 1990,6,419-431.

[2] PlumLee KH, Galey FD. Neurotoxic Mycotoxins: A review of fungal toxins that cause neurological disease in large animals. *Journal of Veterinary Internal Medicine*, 1994,8,49-54.

[3] PlumLee KH. Mycotoxins. In: Robinson NE, editor. *Current Therapy in Equine Medicine*, 4th ed. Philadelphia (PA): WB Saunders; 1997; 668-670.

[4] Osweiler GD. Mycotoxins. *Veterinary Clinics of North America: Equine Practice* 2001,17,547-565.

[5] Cheeke PR, Shull LF. Mycotoxins associated with forages. In: *Natural Toxicants in Feeds, Forages and Poisonous Plants*. Ed. 2. Westport (CT): AVI Publishing; 1998. p. 243-273.

[6] Cheeke PR, Shull LF. Mycotoxins in cereal grains and protein supplements. In: *Natural Toxicants in Feeds, Forages and Poisonous Plants*. Ed. 2. Westport (CT): AVI Publishing; 1998. p. 243-273.

[7] Hollinger K, Ekperigin HE. Mycotoxicosis in food producing animals. *Veterinary Clinics of North America: Food Animal Practice,* 1999,15,133-165.

[8] Ominski, KH, Marquardt, RR, Sinha RN, Abramson D. Ecological aspects of growth and mycotoxin production in storage fungi. In: Miller, JD, Trenholm, HL, editors. *Mycotoxins in grain: Compounds other than aflatoxin*. St. Paul: Eagan Press; 1994. p. 287-312.

[9] Wilson DM, Abramson D. Mycotoxins. In: Sauer DB, editor. Storage of cereal grains and their products. St. Paul, MN; *American Association of Cereal Chemists,* 1992, 341-391.

[10]Osweiler, GD. Mycotoxins: Contemporary issues of food animal health and productivity. *Veterinary Clinics of North America: Food Animal Practice,* 2000, 16, 511-529.

[11]Osweiler GD. Mycotoxins. In: Straw BE, Zimmerman JJ, D'Allaire SC, Taylor DJ, editors. *Diseases of Swine*, 9th ed. Ames (IA): Blackwell Publishing; 2006. pp. 915-929.

[12]Osweiler GD, Carson TL, Buck WB, Van Gelder GA. Mycotoxicoses. *Clinical and diagnostic veterinary toxicology*, 3rd edition. Dubuque (IA): Kendall Hunt, 1985. p. 409-442.

[13]Pier AC. Major biological consequences of aflatoxicosis in animal production. *Journal of Animal Science,* 1992, 70, 3964-3967.

[14]Vesonder R, Haliburton J, Stubblefield R, Gilmore W, Peterson S. *Aspergillus flavus* and aflatoxins B_1, B_2, and M_1 in corn associated with equine death. *Archives of Environmental Contamination and Toxicology*, 1991,20,151-153.

[15]Lee LS, Parrish FW, Jacks TJ. Substrate depletion during formation of aflatoxins and kojic acid on corn inoculated with *Aspergillus flavus. Mycopathologia*, 1986,93,105-107.

[16]Angsubhakorn S, Poomvises P, Romruen K, Newberne PM. Aflatoxicosis in horses. *Journal of the American Veterinary Medical Association,* 1981,178,274-278.

[17]Asquith RL, Edds GT, Aller WW, Bortell R. Plasma concentrations of iditol dehydrogenase (sorbitol dehydrogenase) in ponies treated with aflatoxin B_1. *American Journal of Veterinary Research*, 1980,41,925-927.

[18]Applebaum RS, Brackett RE, Wiseman DW, Marth EH. Responses of dairy cows to dietary aflatoxin, feed intake and yield, toxin content and milk quality of cows treated with pure and impure aflatoxin. *Journal of Dairy Science*. 1982,65(8),1503-08.

[19]Coppock RW, Christian RG. Aflatoxins. In: Gupta RC, editor. *Veterinary Toxicology*. St. Louis: Elsevier, Inc.; 2007. p. 939-950.

[20]Beasley VR. Equine leukoencephalomalacia/hepatosis and stachybotryotoxicosis. In: Robinson NE, editor. *Current Therapy in Equine Medicine*. 3rd ed. Philadelphia: WB Saunders; 1992. p. 377-380.

[21]Diaz GJ, Boermans HJ. Fumonisin toxicosis in domestic animals: A review. *Veterinary & Human Toxicology*, 1994,36,548-555.

[22]Marasas WFO, Kellerman TS, Gelderblom WCA, Coetzer JA, Thiel PG, van der Lugt JJ. Leukoencephalomalacia in a horse induced by fumonisin B_1 isolated from *Fusarium moniliforme*. *Onderstepoort Journal of Veterinary Research,* 1988,55,197-203.

[23]Nelson PE, Plattner RD, Shackelford DD. Desjardins AE. Fumonisin B1 production by Fusarium species other than *F. moniliforme* in section *Liseola* and by some related species. *Applied and Environmental Microbiology*,1992,58,984-989.

[24]Ross PF, Ledet AE, Owens DL Rice LG, Nelson HA, Osweiler GD, Wilson TM. Experimental equine leukoencephalomalacia, toxic hepatosis, and encephalopathy caused by corn naturally contaminated with fumonisins. *Journal of Veterinary Diagnostic Investigation.* 1993,5,69-74.

[25]Vesonder R, Haliburton J, Golinski P. Toxicity of field samples and *Fusarium moniliforme* from feed associated with equine-leucoencephalomalacia. *Archives of Environmental Contamination and Toxicology*, 1989,18,439-442.

[26]Bezuidenhout SC, Gelderblom WCA, Gorst-Allman CP, Horak RM, Marasas WFO, Spiteller G, Vleggaar R. Structure elucidation of the fumonisins, mycotoxins from *Fusarium moniliforme*. *Journal of the Chemical Society, Chemical Communications articles*, 1988,743-5.

[27]Masri MD. Mycotoxic encephalomalacia. In: Robinson NE, editor. *Current Therapy in Equine Medicine*, 3rd ed. Philadelphia (PA): WB Saunders; 1992. p. 558.

[28]Ross PF, Rice LG, Reagor JC, Osweiler GD, Wilson TM, Nelson HA, Owens DL, Plattner RD, Harlin KA, Richard JL, Colvin BM, Banton MI. Fumonisin B_1 concentrations in feeds from 45 confirmed equine leukoencephalomalacla cases. *Journal of Veterinary Diagnostic Investigation*, 1991,3,238-241.

[29]Smith GW. Fumonisins. In: Gupta RC, editor. *Veterinary Toxicology*. St Louis: Elsevier; 2007. p. 983 – 996.

[30]Trenholm HL, Prelusky DB, Young JC, Miller JD. A practical guide to the prevention of fusarium mycotoxins in grain and animal feedstuffs. *Archives of Environmental Contamination and Toxicology*, 1989,18,443-451.

[31]Mostrom MS, Raisbeck MF. Trichothecenes. In: Gupta RC, editor. *Veterinary Toxicology*. St Louis: Elsevier; 2007. p. 951-976.

[32]Wolf-Hall CE, Hanna MA, Bullerman LB. Stability of deoxynivalenol in heat-treated foods. *Journal of Food Protection*,1999,62,962-964.

[33]Wolf CE, Bullerman LB. Heat and pH alter the concentration of deoxynivalenol in an aqueous environment. *Journal of Food Protection*, 1998,61,365-367. Erratum in: *Journal of Food Protection*, 1998,61,667.

[34]Croft WA, Jarvis BB, Yatawara CS. Airborne outbreak of trichothecene mycotoxicosis. *Atmospheric Environment*, 1986, 20,549-552

[35]Coppock R, Sziwenka MM: Stachybotryotoxins. In: PlumLee KH, editor. *Clinical Veterinary Toxicology*, St. Louis (MO): Mosby; 2004. pp. 268-270.

[36]Jarvis BB, Lee YW, Comezoglu SN, Yatawara CS. Trichothecenes produced by *Stachybotrys atra* from Eastern Europe. *Applied and Environmental Microbiology,* 1986,51(5),915-8.

[37]Gallagher RT, Campbell AG, Hawkes AD, Holland PT, McGaveston DA, Pansier EA. Ryegrass staggers: the presence of lolitrem neurotoxins in perennial ryegrass seed. *New Zealand Veterinary Journal*, 1982,30,183-184.

[38]Crump MH. Slaframine (slobber factor) toxicosis. *Journal of the American Veterinary Medical Association,* 1973,163,1300-1302.

[39]Hagler WM, Behlow R. Salivary syndrome in horses: Identification of slaframine in red clover hay. *Applied and Environmental Microbiology,* 1981,42,1067-1073.

[40]Hagler WM, Croon WJ. Slaframine: occurrence, chemistry, and physiological activity. In: Cheeke PR, editor. *Toxicants of Plant Origin.* Vol 1: Alkaloids. Boca Raton (FL): CRC Press; 1989. p. 257-279.

[41]Harvey RB, Edrington TS, Kubena LF, Elissable MH, Rottinghaus GE. Influence of aflatoxin and fumonisin B_1-containing culture material on growing barrows. *American Journal of Veterinary Research,* 1995,56,1668-1672.

[42]Harvey RB, Kubena LF, Huff WE, Rottinghaus G. Effects of aflatoxin, deoxynivalenol, and their combinations in the diets of growing pigs. *American Journal of Veterinary Research,* 1989,50,602-607.

[43]Meerdink GA. Aflatoxins. In: PlumLee KH, editors. *Clinical Veterinary Toxicology*. St. Louis: Elsevier – Mosby; 2004. p. 231-235.

[44]Aller WW, Edds GT, Asquith RL. Effects of aflatoxins in young ponies. *American Journal of Veterinary Research,* 1981,42, 2162-2164.

[45]Asquith RL. Equine aflatoxicosis: A brief review. In: Lacey JL, editor. *Trichothecenes and other mycotoxins*. New York: John Wiley & Sons; 1985. p. 487.

[46]Cysewski S, Pier AC, Baetz A, Cheville NF. Experimental equine aflatoxicosis. *Toxicology and Applied Pharmacology,* 1982,65,354-365.

[47]Green HT, Oehme FW. A possible case of equine aflatoxicosis. *Clinical Toxicology,* 1976,9,251-252.

[48]McGavin MD, Knake R. Hepatic midzonal necrosis in a pig fed aflatoxin and a horse fed moldy hay. *Veterinary Pathology,* 1977,14,182-187.

[49]Bortel LR. Clinical chemistry and hematology of weanling ponies exposed to aflatoxin B_1 [Master's thesis]. Gainesville (FL): University of Florida, Gainesville; 1981.

[50]Schurg WA, Noon TN. Experimentally induced Aflatoxicosis in mature horses fed naturally contaminated cotton products. Proceedings of the 6th *Equine Nutrition and Physiology Symposium*; College Station, TX, 1979, p. 72.

[51]Park DL, Lee LS, Price RL, Pohland AE. Review of the decontamination of aflatoxins by ammoniation: Current status and regulation. *Journal of the Association of Official Analytical Chemists,* 1988,71,685-703.

[52]Goel S, Schumacher J, Lenz SD, Kemppainen B. Effects of *Fusarium moniliforme* isolates on tissue and serum sphingolipid concentrations in horses. *Veterinary & Human Toxicology,* 1996,38,265-270.

[53]Kwon O, Sandberg J, Slikker W. Effects of fumonisin B_1 treatment on blood-brain barrier transfer in developing rats. *Neurotoxicology and Teratology,* 1997,19,151-155.

[54]Ramasamy S, Wang E, Hennig B, Merrill AH Jr. Fumonisin B_1 alters sphingolipid metabolism and disrupts the barrier function of endothelial cells in culture. *Toxicology and Applied Pharmacology,*1995,133,343-348.

[55]Vos KA, Smith GW, Haschek WM. Fumonisins: Toxicokinetics, mechanism of action and toxicity. *Animal Feed Science and Technology,*2007,137,299-325.

[56]Smith GW, Constable PD, Foreman H, Eppley RM, Waggoner AL, Tumbleson ME, Haschek WM. Cardiovascular changes associated with intravenous administration of fumonisin B_1 in horses. *American Journal of Veterinary Research,* 2002,63,538-45.

[57]Wilson TM, Ross PF, Owens DL. Experimental reproduction of ELEM. A study to determine the minimum toxic dose in ponies. *Mycopathologia,* 1992,117,115-20.

[58]Ross PF, Rice LG, Osweiler GD, Nelson PE, Richard JL, Wilson TM. A review and update of animal toxicoses associated with fumonisin-contaminated feeds and production of fumonisins by *Fusarium* isolates. *Mycopathologia*, 1992,117,109-114.

[59]Wilson TM, Ross PF, Rice LG, Osweiler GD, Nelson HA, Owens DL, Plattner RD, Reggiardo C, Noon TH, Pickrell JW. Fumonisin B_1 levels associated with an epizootic of equine leukoencephalomalacia. *The Journal of Veterinary Diagnostic Investigation,* 1990,2,213-216.

[60]McCue PM. Equine leukoencephalomalacia. *Compendium on Continuing Education for the Practicing Veterinarian,* 1989,11,646-650.

[61]U.S. Department of Health and Human Services, Food and Drug Administration, Center for Food Safety and Nutrition, Center for Veterinary Medicine. June, 2000. Available from: http://vm.cfsan.fda.gov/~dms/fumongui.htmL.

[62]Jouany JP. Methods for preventing, decontaminating and minimizing the toxicity of mycotoxins in feeds. *Animal Feed Science and Technology,* 2007,137,342-362.

[63]Weaver GA. Experimental mycotoxicoses in food producing animals [dissertation]. University of Minnesota; 1981.

[64]Johnson PJ, Casteel SW, Messer NT. Effect of feeding deoxynivalenol (vomitoxin)-contaminated barley to horses. *The Journal of Veterinary Diagnostic Investigation,* 1997,219-221.

[65]Gonzalez-Cabrero J, Edigo J, Barat A. Detection and characterization of circulating and glomerular immune complexes in experimental IgA nephropathy. *Immunology*, 1990, 70,296-302.

[66]Raymond SL, Smith TK, Swamy HVLN. Effects of feeding a blend of grains naturally contaminated with *Fusarium* mycotoxins on feed intake, serum chemistry, and hematology of horses, and the efficacy of a polymeric glucomannan mycotoxin adsorbent. *Journal of Animal Science*, 2003,81,2123-2130.

[67]Raymond SL, Smith TK, Swamy HVLN. Effects of feeding a blend of grains naturally contaminated with *Fusarium* mycotoxins on feed intake, metabolism and indices of athletic performance of exercised horses. *Journal of Animal Science*, 2005,83,1267-1273.

[68]Leung MC, Smith TK, Karrow NA, Boermans HJ. Effects of food borne Fusarium mycotoxins with and without a polymeric glucomannan mycotoxin adsorbent on food intake and nutrient digestibility, body weight, and physical and clinicopathologic variables of mature dogs. *American Journal of Veterinary Research,* 2007,68(10),1122-9.

[69]DiMenna ME, Mortimer PH, Prestidge RA, Hawkes A, Sprosen J. Lolitrem B concentrations, counts of *Acremonium lolii* hyphae, and the incidence of ryegrass staggers in lambs on plots of *A. lolii*-infected perennial ryegrass. *New Zealand Journal of Agricultural Research,* 1992,35,211-217.

[70]Blodgett DJ. Fescue Toxicosis. In: Galey FD, editor. *The Veterinary Clinics of North America: Equine Practice.* Philadelphia: WB Saunders. 2001,17,567-577.

[71]Blodgett DJ. Fescue Toxicosis. In: Gupta RC, editor. *Veterinary toxicology.* St Louis: Elsevier; 2007. p. 907-914.

[72]Evans TJ, Rottinghaus GE, Casteel SW. Ergot. In: PlumLee KH, editor. *Clinical veterinary toxicology*. St. Louis: Mosby Inc.; 2004. p. 239-43.

[73]Wright R. Molds, mycotoxins and their effect on horses [online]. January 1, 2005. Available from http://www.omafra.gov.on.ca/english/livestock/horses/facts/info mycotoxin.htm.

[74]Riet-Correa F, Mendez MC, Bergamo PN, Flores WN. Agalactia, reproductive problems and neonatal mortality in horses associated with the ingestion of *Claviceps purpurea*. *Australian Veterinary Journal,* 1988,65(6),192-193.

[75]Evans TJ, Rottinghaus GE, Casteel SW. Fescue. In: PlumLee KH, editor. *Clinical Veterinary Toxicology*. St. Louis: Mosby Inc.; 2004. p. 243-250.

[76]Cross DL, Redmond LM, Strickland JR. Equine fescue toxicosis: signs and solutions. *Journal of Animal Science,* 1995,73(3),899-908.

[77]Putnam MR, Bransby DI, Schumacher J, Boosinger TR, Bush L, Shelby RA, Vaughn JT, Ball D, Brendemuehl JP. Effects of the fungal endophyte *Acremonium coenophialum* in fescue on pregnant mares and foal viability. *American Journal of Veterinary Research*, 1991,52, 2071-2074.

[78]Cross DL. Fescue toxicosis in horses. In: Bacon CW, Hill NS, editors. *Neotyphodium/Grass Interactions*. New York: Plenum Press; 1997. p. 289-309.

[79]Brendemuehl JP, Williams MA, Boosinger TR, Ruffin DC. Plasma progestagen, tri-iodothyronine, and cortisol concentrations in postdate gestation foals exposed *in utero* to the tall fescue endophyte *Acremonium coenophialum*. In: Equine Reproduction VI. *Biology of Reproduction Mono 1,* 1995, 53-59.

[80]Sockett DC, Baker, JC and Stowe CM. Slaframine (*Rhizoctonia leguminocola)* intoxication in horses. *Journal of the American Veterinary Medical Association*, 1982,181,606.

In: Mycotoxicoses in Animals Economically Important
Editors: Edlyane Gonçalez, et. al. pp. 35-66
ISBN: 978-1-61668-195-1

Chapter 2

THE SIGNIFICANCE OF FEED-BORNE MYCOTOXINS IN RUMINANT NUTRITION

***Trevor K. Smith*[1] *and Sofya N. Korosteleva*[2]**
[1] Professor, Department of Animal and Poultry Science,
University of Guelph, Guelph, Ontario, Canada
[2] Ruminant Nutritionist,
Nutreco Canada Inc., St. Mary's, Ontario, Canada

ABSTRACT

It has been estimated that 25% of crops worldwide are contaminated with mycotoxins annually. Mycotoxins lead to significant economic losses due to effects on livestock, losses in crops and regulatory/research costs. Mycotoxin screening of feed samples submitted by North Carolina farmers over a nine year period indicated that up to 70% of samples were measurably contaminated with one or more of 5 analyzed mycotoxins. A recent study from the Netherlands included the screening of 169 feed samples from 24 dairy farms for 20 mycotoxins. Corn silage was the most commonly contaminated feedstuff. Mycotoxins were present in up to 54% of samples. Such analysis of mycotoxin contamination of feed is usually not complete because conjugated mycotoxins are not detected. These metabolites are the product of plant metabolism and are usually found as glycosides. Conjugated mycotoxins are not detected by routine laboratory analyses. It has been shown that up to 70% of DON can be present in non-detectable conjugated forms. Conjugated mycotoxins are thought to be degraded to free mycotoxins *in vivo* during digestion. Symptoms of mycotoxicosis in dairy cattle vary depending on the mycotoxins involved and their interaction with other stress factors. Highly stressed animals, such as fresh cows, are the most sensitive to mycotoxicoses because they suffer from a degree of immunosuppression. Symptoms of mycotoxicoses may be non-specific including reduced production, feed consumption and impaired reproduction. Transfer of aflatoxins into milk of lactating animals is a major public health issue due to the carcinogenic nature of these compounds. The rate of transfer of mycotoxins from blood into milk can be affected by systemic or local disease leading to impaired transport barriers. The level of milk production will also influence aflatoxin residues in milk. Rumen function has been shown to be negatively affected by feed-borne mycotoxins, suggesting a significant role of the rumen in handling mycotoxin exposure, biotransformation and excretion of metabolites. Mycotoxins undergo microbial biotransformation in the rumen and this can lead to production of metabolites which may be more or less toxic. The capacity of the rumen to minimize the harmful effects of mycotoxin exposure will depend on multiple factors including type of diet, rumen health and bioavailability of mycotoxins in a feed. Immunity and rumen health are the factors most affected by mycotoxin exposure. Good feed management should be practiced to minimize diet contamination. Maximizing immunity and rumen health will minimize the harmful effects of mycotoxin exposure.

INTRODUCTION

Mycotoxin occurrence and concentrations vary by year. This is expected because of the annual variation in weather conditions and plant stresses which are known to affect mycotoxin formation [1]. Mycotoxins occur frequently in a variety of feedstuffs and are routinely fed to animals. Sometimes, mycotoxins occur at concentrations high enough to cause major losses in health and performance of animals. A more common scenario, however, is to find mycotoxins at lower levels interacting with other stressors to cause subclinical losses in performance, increases in the incidence of disease and reduced reproductive performance. To the animal producer, these subclinical losses are of greater economic importance than losses from acute effects but they are even more difficult to diagnose [2].

CONTAMINATION OF CATTLE FEED WITH MYCOTOXINS

The contamination of animal feed with mycotoxins is a worldwide problem in animal production. The complex diet of ruminants, consisting of forages, concentrates, and silages, can be a source of diverse mixtures of mycotoxins that contaminate individual feed components. The major classes of mycotoxins of concern in feedstuffs include aflatoxins, trichothecenes, zearalenone, fumonisins, ochratoxin and ergot alkaloids [3]. Mycotoxins affect cattle health, reproduction and performance and mycotoxin residues in milk can pose a hazard to human health. A number of mycotoxins are inactivated by rumen microflora and are transformed to less toxic metabolites, other mycotoxins may be unchanged or are converted into metabolites that have increased toxicity.

Mycotoxins originating from pre-harvest contamination of forages that are ensiled remain unaffected by the ensiling process. Preserved feed such as silage, hay, and straw can be affected by storage molds which can flourish depending on feed storage management rather than on climatic conditions. After prolonged storage periods, silage can be spoiled by a variety of fungal species, which are acid-tolerant and micro-aerobic. Mycological investigations identified *Penicillium* spp. (*P. roqueforti* and *P. paneum*), *Aspergillium* spp. (*A. fumigatus* and *A. flavus*) [4], *Monascus* spp. [5], and *Byssochlamys nivea* [6] as the most prevalent fungal species in silage [7]. Mycotoxins produced by these fungi include patulin, mycophenolic acid, penicillic acid, roquefortines, marcfortine A, andrastin A, gliotoxin, and toxins of the verruculogen/ fumitremorgen group [8, 9].

Mycotoxin screening of feed samples submitted by North Carolina farmers over a nine year period indicated that up to 70% of samples were contaminated to some degree with one or more of 5 analyzed mycotoxins (aflatoxin (AF), deoxynivalenol (DON), zearalenone (ZON), T-2 toxin and fumonisin) [10]. The most frequently contaminated feedstuffs were corn silage and corn grain.

The screening results of 169 feed samples from 24 Dutch dairy farms for 20 mycotoxins have recently been published [11]. Corn silage was the most contaminated feedstuff. Mycotoxins were present in up to 54% of samples. The incidence of DON and ZON in silage, compound feed, and feed commodity samples was 38 to 54% and 17 to 38% respectively. DON and ZON were not detected in ensiled by-products. Roquefortine C and mycophenolic acid were only detected in silage and ensiled by-product samples with an incidence of 7 to 19%. Fumonisins B_1 and B_2 were detected in 2 compound feed samples and one feed commodity sample.

Average concentrations of DON, ZON, roquefortine C, and mycophenolic acid in complete diets were 273, 28, 114, and 54 μg/kg, respectively. Maximum concentrations were 969, 203, 2,211, and 1,840 μg/kg, respectively. Calculated average daily intakes of these mycotoxins were 5.0, 0.5, 2.0, and 0.9 mg/animal, respectively, and maximum daily intakes 19.3, 3.5, 38.9, and 32.3 mg/animal, respectively. Corn silage was the major dietary source of all 4 of these mycotoxins. Extremely high concentrations of roquefortine C and mycophenolic acid (up to 45 and 25 mg/kg, respectively) were detected in visibly molded areas in surface layers of corn silage.

Prevalence of roquefortine C and mycophenolic acid in silage samples of 56% and 32%, respectively, have been reported in surveys in Germany [12,13].

Pasture grasses have been found to be contaminated with the *Neotyphodium* toxins of the lolitrem paxilline group and ergovaline, as well as other ergot alkaloids [7]. The level of contamination of grasses shows significant geographical differences [14] and is gaining increasing attention. Recently, ergot alkaloids, originating from *Claviceps* and other endophytes, have been found with increasing frequency in pasture grasses of Northern Europe at levels that are toxic to animals [15, 16]. Tall fescue (*Festuca arundinacea*) is the major pasture grass in large parts of the United States. Tall fescue is commonly infected by endophytes (*Neothyphodium* spp. *and Claviceps* spp.) that produce a range of alkaloids frequently detected. Ergovaline is commonly measured as a marker toxin [17].

The most frequently detected classes of mycotoxins in contaminated grains and protein byproducts are the trichothecenes, fumonisins and zearalenone. Cereal grains are frequently contaminated with deoxynivalenol, ochratoxins (OT), and ergot alkaloids [18-21].

Naturally contaminated feed usually contains combinations of mycotoxins which can have additive or synergistic effects and cause mycotoxicoses even if concentrations of the major mycotoxin are low [22]. This may explain the variety of effects observed in field studies of *Fusarium* mycotoxins as well as in trials with naturally contaminated feed compared with experiments utilizing pure mycotoxins [23-29].

Conjugated mycotoxins, in which the toxin is usually bound to a more polar substance such as glucose, are referred to as "masked" mycotoxins. These mycotoxins can escape routine detection methods but can be released after hydrolysis *in vivo*. Berthiller et al. [30] reported the natural occurrence of a glycoside of deoxynivalenol in *Fusarium*-infected wheat and corn to be up to 12% of DON content. More recent research on conjugated DON in barley indicated the release of an additional 58% DON following acid hydrolysis [31]. Sasanya et al. [32] found free and conjugated DON in wheat and reported recovering of

DON-3-glucoside at 70.0%. It has also been reported that conjugated mycotoxins are hydrolyzed during digestion and can contribute to mycotoxicoses [33].

Mycotoxin Degradation in the Rumen

Evidence of ruminal microbial degradation of mycotoxins has been demonstrated in isolated cultures of rumen content [34-37]. Interest in this area was stimulated by early studies revealing the greater resistance of ruminants to the negative effects of some mycotoxins [38]. Recently the rumen stability of different mycotoxins and their impact on rumen health was reviewed by Jouany and Diaz [39].

It was demonstrated that 90–100% of the metabolism of OT, ZON, T-2 toxin, and diacetoxyscirpenol (DAS) was achieved by rumen protozoa and, therefore, they were considered to be the most important ruminal microbial population contributing to mycotoxin biodegradation. Considerably lower metabolism of the same mycotoxins was noted in cultures containing only rumen bacteria. Neither AF nor DON was metabolized by rumen microorganisms [37].

Contradictory results have been reported regarding rumen biodegradation of aflatoxin B_1. Engel and Hagemeiser [40] found that 42% of AFB_1 was degraded when incubated *in vitro* with rumen fluid. Other researches found insignificant or no degradation of AFB_1 [37], [41, 42]. Some aflatoxin metabolites such as AFM_1 and aflatoxicol have been detected in rumen contents [43]. *In vitro* studies employing radioactive labeled AFB_1 showed that aflatoxicol produced by rumen microbes accounted for only 0.6% of initial radioactivity [44]. Since there is only minor degradation of AFB_1 in the rumen and the toxicity of the main rumen metabolite (aflatoxicol) is similar to that of the parent molecule, it can be concluded that the ruminants are not effectively protected against aflatoxin toxicity [39]. Upadhaya et al. [45] found aflatoxin degradability was higher in goats compared to steers. It was also noticed that degradability was higher with the feeding of timothy hay as opposed to rice straw, although individual variation was seen.

In vitro studies indicated that more than 90% of zearalenone is degraded to α-zearalenol by rumen microorganisms [37]. This metabolite is considered to be about 4-times more oestrogenic than the parent toxin [46]. Lesser production of β-zearalenol, which has a low affinity to oestrogenic receptors but is toxic to endometrial cells, was also reported [147]. Protozoa have been reported to be about nine times more active in ZON degradation than bacteria [37].

De-epoxidation and de-esterization of thrichothecenes in the rumen results in production of several metabolites [48, 49]. Rumen microflora de-epoxydaze DON to DOM-1 [50, 51]. Since the epoxide group on the trichothecene nucleus causes cytotoxicity and vomiting [52], biotransformation to a diene group reduces the toxicity of DON in ruminants. *In vivo* studies were conducted by Danicke et al. [27] on cows fed 8.05 and 7.15 mgDON/kg feed and 0.26 and 0.1 mgZON/kg contaminated wheat. It was observed that DON was nearly completely metabolized to de-epoxy-DON and the flow at the duodenum ranged between 4% and 28% of DON intake. The ZON metabolites α-zearalenol (ZOL) and β-ZOL were recovered at the duodenum as addition to the parent toxin ZON ranging from of 43% to 132% of ZON intake.

T-2 toxin is deacetylated to HT-2 and further to T-2 triol in rumen fluid [53, 54]. Westlake et al. [42] reported that toxin-degrading activity was higher in protozoa than in bacteria. Ninety percent of T-2 roxin and DAS degradation is achieved by protozoa [55].

It has been observed that Fumonisin B_1 is poorly metabolized in the rumen. Only 10-18% of the initial amount added to rumen fluid *in vitro* disappeared after 72 hr of incubation [56, 57]. Rumen metabolites produced from FB_1 remain to be characterized [56].

Ochratoxin A (OTA) is converted in the rumen into phenylalanine and OTA-α [39]. OTA-α is less toxic than the parent compound, but retain some genotoxicity [58]. OTA also can be esterified to produce ochratoxin C which has similar toxicity [59]. Up to 60% of dietary OTA was reported to be degraded to OTA- α during *in vitro* incubation with rumen fluid [34]. Protozoa are considered to be the main microorganisms involved in the conversion of OTA to OTA-α in the rumen [37]. This could explain why ruminants are less sensitive to acute ochratoxicosis than monogastrics [35]. Muller [60] estimated from *in vitro* data that an adult cow is able to degrade 33 to 72 mg of OTA per day. Sheep have been shown to be sensitive to contaminated concentrate feeds when dosed at 20 mgOTA/kg BW (6-14 mg/day). Such differences in rumen OTA detoxification could be due to the accessibility of microbes to the toxin [61]. The use of crystalline OTA may have enhanced microbial access and thus breakdown whereas OTA fed as part of a naturally contaminated feed may be partially protected by the feed matrix. The type of diet, which influences rumen microbial composition and activity, may also affect the extent of OTA degradation. Metabolism of OTA in the rumen has been reported much lower in grain-fed ruminants [62, 63]. The respective ruminal half-lives were 2.7 or 0.63 hrs for OTA and 1.9 and 0.9 hrs for OTA- α when animals were fed a grain-based or hay-based diet respectively. The authors calculated that OTA bioavailability was 4.3 times greater in sheep fed grain than in sheep fed hay. Feeding the complete grain diet resulted in a severe pH drop to 5.6. This led to a significant decrease in the number of protozoa and to a lower OTA degradation rate. Muller et al. [64, 65] and Ozpinar et al. [66], however, reported that high a concentrate starch-based diet accelerated OTA degradation. They confirmed that protozoa are more active in metabolizing OTA than bacteria. The capacity of rumen fluid to degrade OTA follows the kinetics of protozoa concentration in the rumen. Ruminal degradation of OTA decreased for the first 1.5 hrs after feeding, and subsequently increased, peaking 5 hrs after exposure [37].

The most prominent class of toxins that is not inactivated by the rumen microflora are the fungal alkaloids and related indole diterpenoids [67]. Intoxication with ergot alkaloids was one of the first mycotoxicoses to be reported in animals.

Capacity of the rumen to inactivate mycotoxins can be influenced by the type of diet and the use of additives and medications. These factors influence microbial composition and population and rumen passage rate. Stress factors such as decreased feed intake in the transition cow also play a role. Bioavailability of natural occurring mycotoxins to rumen microflora might be affected by the protection of the mycotoxin by the matrix of feed. Mycotoxin conjugation, which can increase the time required for degradation and, subsequently, increase the chance of mycotoxins escaping the rumen and becoming available to the animal, can also be a significant factor.

Influence of Mycotoxins on Rumen Health

Rumen function has been shown to be negatively affected by mycotoxins, suggesting a role for the rumen in limiting mycotoxin exposure and promoting biotransformation and excretion of metabolites [68]. The negative effects can include decreased cellulose degradation, altered volatile fatty acid (VFA) production, increased ammonia production, proteolysis and reduced rumen motility [68, 69].

The addition of AFB_1 at 9.5 ng/mL to rumen fluid did not alter *in vitro* digestion of alfalfa or VFA production [5]. Westlake et al. [42], however, measured a large decrease in the digestive capacity of rumen microbes when AFB_1 was added at 1 μg/mL and 10 μg/mL which resulted in 50% and 67% reduction in alfalfa digestibility respectively. Such contrasting results could be explained by differences in mycotoxin concentrations [39]. Compromised ruminal function was reported by number of researchers. Cows exposed to aflatoxins had reduced cellulose digestion, volatile fatty acid production and rumen motility [68], [70, 71]. Cook et al. [68], used radiotelemetry to measure rumen motility and found that AF administration (200–800 μg/kg) slowed motility in a dose-dependent manner.

Postprandial rumen fluid ammonia concentration was consistently higher when wheat naturally contaminated with DON and ZON was fed to dairy cows [27]. The flow of microbial protein and utilizable protein at the duodenum, moreover, were reduced. These results suggest that DON can alter ruminal microbial protein synthesis. A mixture of DON and ZON fed to rumen-cannulated wethers at concentrations of 4.6 and 0.34 μg/kg TMR decreased the *in sacco* rate of straw degradation, whereas no changes occurred when alfalfa was incubated in the rumen [72].

T-2 toxin induced an 83% decrease in ruminal protozoal protein whereas bacterial protein increased by 14% during a 6 hrs incubation period [55]. This negative effect on eucariotic cells is due to T-2 binding to ribosomes and inhibiting protein synthesis [73].

Two rumen bacteria responsible for cellulolysis (*Ruminococcus albus)* and metabolism of hydrogen (*Methanobrevibacter ruminantium*) have been reported to be inhibited by fusaric acid at concentrations as low as 15 μg/mL. DON had no such effect at 100 μg/mL [74].

Fumonisin B_1 seems to have little effect on rumen function. *In vitro* dry matter digestibility of feeds was not significantly changed by FB_1 contamination indicating that microbial efficiency was unaffected by the toxin at dietary concentrations up to 100 mg/kg [57]. No significant effect of FB_1 was observed on VFA production [56].

The toxicity of OTA to rumen microbes has been investigated only for *Butyrivibro fibrisolvens* CE51 [75] as this bacterial species is considered to be involved in the rumen mycotoxin detoxification process [73]. The growth of this bacterium was not affected by the presence of OTA. No significant effect of OTA was observed on feed digestibility in sheep when present in the diet at 5 mg/kg [61].

Patulin produced by *Penicillium, Aspergillus*, and *Byssochlamys* spp. can be commonly found in silages [76]. Patulin inhibits the growth of gram + and gram - bacteria, and some protozoa [77,78]. Escoula [79] showed that patulin had a negative dose-dependant effect on VFA production (especially on acetate production) and microbial protein synthesis, when concentrations increased from 20 to 300 μg/mL of rumen fluid. Tapia et al. [80] observed a dose-dependant suppression of digestion of a mixed diet in an artificial rumen at patulin concentrations 30 mg/kg and 90 mg/kg. VFA production (acetate) was reduced at 90 mg/kg only. Results indicate a negative effect of patulin on cellulose digestion. The outflow of microbial protein was lowered by presence of patulin independent, of dose. The negative effect on the efficiency of microbial protein synthesis (g microbial N/kg fermented OM), however, was reached only at the highest dose of 90 mg/kg.

While rumen flora acts as first line of defense against mycotoxins the negative effect on the microflora influences overall rumen health as well as feed digestion and utilization. The detoxifying capacity of rumen microflora can be reduced under continuous exposure to mycotoxins resulting in passage of mycotoxins through the rumen unchanged with subsequent absorption in the duodenum resulting in negative effects on animal health and performance.

Toxicity Studies in Cattle, Toxicokinetics and Carry Over of Mycotoxins into Milk

Aflatoxins

Aflatoxins are produced by *Aspergillus flavus* and *A. parasiticus*. These are tropical and semi-tropical molds that thrive at high temperatures under both high humidity and drought conditions. Carcinogenic, mutagenic and teratogenic effects of AFB_1 have been reported for several animal species, including humans [81].

Toxicokinetics of AF

Aflatoxins are the most intensively studied mycotoxins in dairy cattle as the residues of aflatoxin M_1 in milk is a public health concern. Some aflatoxin B_1 is degraded in the rumen, following ingestion of aflatoxin-contaminated feeds. This result in the formation of aflatoxicol. The remaining aflatoxin is absorbed in the digestive tract by passive diffusion and is hydroxylated in the liver to aflatoxin M_1 [82]. Aflatoxin M_1 is either conjugated to glucuronic acid and subsequently excreted in bile, or it enters the systemic circulation. Circulating AFM_1 can be excreted in the urine or appear in milk. AFM_1 and to lesser extent AFM_2 are the main aflatoxin metabolites found in milk. Other metabolites such as AFM_4, AFQ_1 and aflatoxicol have also been detected [83].

Effect of AF on Ruminants

Aflatoxins primarily target the liver and the immune system. Symptoms of acute aflatoxicosis includes: inappetance, lethargy, ataxia, rough hair coat, and enlarged fatty liver. Chronic aflatoxin exposure is characterized by liver cell injury, a fatty liver syndrome, poor feed conversion, and a significant reduction in milk yield [68], [84-86]. Increasing AF in cattle feed to levels 10, 26, 56.4, 81.1, and 108.5 μg/kg has been shown to significantly reduce feed intake at each level in a dose-dependent manner [87]. High-yielding dairy cows are considered to be more sensitive to aflatoxins than fattening cattle [88]. Production and health of dairy herds may be affected at dietary aflatoxin levels above 100 μg/kg which is considerably higher than the amount that results in illegal milk residues [89, 90]. Guthrie [91] observed a decline in reproductive efficiency in lactating dairy cows under field conditions when they were consuming 120 μg/kg aflatoxin. Milk production increased over 25% when cows were switched to an aflatoxin free diet. Applebaum et al. [88] showed that impure aflatoxin produced by culture reduced production while equivalent amounts of pure aflatoxin did not.

Aflatoxin has also been shown to lower resistance to diseases and to interfere with vaccine induced immunity in livestock [71]. Several mechanisms of bovine immunosuppression by AFB_1 have been demonstrated *in vitro*. Paul et al. [92] observed that AFB_1 suppressed mitogen-induced stimulation of peripheral lymphocytes. In another study, AFB_1 was shown to inhibit bovine lymphocyte blastogenesis [93].

AFB_1 at 600 μg/kg was shown to depress feed efficiency and rate of gain in steers in a 155-day feeding trial [70][82]. Decreased feed efficiency in cattle has been attributed to compromised ruminal function caused by reductions in cellulose digestion, volatile fatty acid production and rumen motility [68], [70, 71]. Garrett et al. [94] showed reduced weight gain and feed intake of beef cattle fed diets containing 700 μg aflatoxin/kg feed. It has been pointed by some researchers that increases in liver weights can result from lower levels of aflatoxin exposure and this can be a benchmark for toxicity in beef cattle [95].

Liver is a site of high zinc metabolic activity and it is also the organ most affected by aflatoxins. Studies on zinc metabolism were performed on dairy calves fed 5 mg/kg of aflatoxin [96]. Zinc concentrations were reduced in liver and increased in kidney and three sections of the small intestine.

Several studies have shown the sensitivity of sheep to AF. Harvey et al. [97] showed that feeding diets contaminated with several aflatoxins to ewe lambs (2.5 mg/kg or 5.0 mg/kg of feed for 35 days) resulted in hepatotoxicity. In another study lambs fed AF at 2.5 mg/kg for 21 days showed symptoms of clinical aflatoxicosis including hepatic and nephritic lesions, altered mineral metabolism, and increased size and weight of the liver and kidney [98]. In another study with the same exposure to AF (2.5 mg/kg of feed) plasma mineral concentrations were determined on day 1, 2, 4 and 8 following the initial exposure [99]. On day 4 of intoxication, significant reductions in plasma concentrations were detected for Ca (2.39 μM vs. 2.06 μM), P (2.95 μM vs. 2.50 μM), Mg (0.88 μM vs. 0.77 μM), K (4.40 μM vs. 3.81 mM), and Zn (13.2 μM vs. 11.6 μM). The resulting mineral deficiencies due to aflatoxicosis were attributed to lower feed intake and to liver and kidney malfunctions resulting from AF intoxication. Exposure of lambs to AF (2.5 mg/kg of feed for 3 weeks) revealed changes in extrinsic coagulation factors as determined by increased fibrinogen concentration [100]. In a study on 5-month-old lambs fed feed contaminated with 2 mg/kg AF for 37 days, average daily gain after 35 days of feeding was significantly reduced from 125 g to 79 g and the exposed lambs showed decreased cellular immunity [101]. After allowing for a clearance period of 30 days, the AF-exposed lambs had average daily gain and cellular immunity similar to those for the controls. Mechanisms for cellular immune response to AF in sheep have not been elucidated.

The US Food and Drug Administration limits aflatoxin to no more than 200 μg/kg in feed for breeding cattle, 300 μg/kg for beef finishing cattle, 20 μg/kg in lactating dairy feeds and to 0.5 μg/kg in milk. Milk aflatoxin concentrations have been estimated around 1.7% of the aflatoxin concentration in the total ration dry matter [102]. Cows consuming diets containing 30 μg AF/kg can produce milk containing aflatoxin residues above the FDA maximum allowed level of 0.5 μg/kg. In Europe, the regulatory levels of aflatoxin are 20 μg/kg for dairy feeds and 0.05 μg/kg in milk. An illegal milk residue can occur, therefore, when feed contains more than 3 μg/kg of aflatoxin [95].

Transfer of AF into Milk

Transfer rate of aflatoxin to milk varies between 1% and 6% of dietary contamination [103]. The extent of transfer from feed to milk is influenced by various nutritional and physiological factors, including feeding regimes, rate of ingestion, rate of digestion, health of the animal, hepatic biotransformation capacity, and actual milk production [7]. In high-yielding cows, the consumption of significantly higher amounts of concentrated feeds might result in carry-over percentages as high as 6.2% [104].

In the study by Applebaum et al. [88], 10 ruminally-cannulated lactating Holstein cows were given AFB_1 (13 mg per cow daily) via the rumen orifice for 7 days. Levels of AFM_1 in milk ranged from 1.05 to 10.58 ng/l. The AFB_1-treated cows also had a significant reduction in milk yield. In another study [104] the carry-over rate was shown to be higher (6.2 vs. 1.8) in early lactation (2–4 week) when compared with late lactation (34–36week). A survey of milk samples from 132 French farms has been conducted [105]. AFM_1 was found in 3 out of 264 samples at levels 26 ng/L or less. These are below the European legislation limit of 50 ng/L. Traces of AFM_1 (less than 8 ng/L) were also found in 6 other samples.

Studies of dairy ewes have indicated that concentrations of AFM_1 in milk were linearly affected by the AFB_1 dose [106]. The carry over was lower than reported for dairy cattle and goats, suggesting a better ability of sheep to degrade AFB_1.

Deoxynivalenol

Deoxynivalenol (DON, vomitoxin), a trichothecene mycotoxin produced by *Fusarium* fungi has been shown to inhibit cellular protein synthesis.

Toxicokinetics of DON

Prelusky et al. [107] reported that DON administered as a single oral dose to dairy cows at levels of 1.9 mg/kg b.w. resulted in less than 1% systemic absorption of the parent toxin. Maximum blood levels in 2 dosed cows reached 200 and 90 ng/mL serum, occurring at 4.7 and 3.5 hrs following dosing, respectively. With an elimination half-life ($t1/2$) of approximately 4 hrs, however, 24 hrs was required for serum DON/metabolite levels to drop below detectable concentrations (2 ng/mL). It was estimated that 24–46% of DON in serum existed as a β-glucuronide conjugate. The de-epoxide metabolite DOM-1 was not analyzed. Free and conjugated DON was also present in milk, but only in extremely low amounts (less than 4 ng/mL) were detected. Côté et al. [108] reported on the excretion and metabolism of DON fed in a diet (average 66 mg/kg) in dairy cows exposed for 5 days. Unconjugated DOM-1 was found in milk at a concentration of 26 ng/mL. Approximately 20% of the consumed toxin was recovered in the urine (~11%) and feces (~9%) as DON (4%) and DOM-1 (96%). It was further noted that enzymatic treatment of the urine increased DON and DOM-1 levels 1.6- to 3- and 7- to 15-fold, respectively, indicating that a significant proportion (>70%) of ingested DON was eliminated in the urine as the DOM-1 conjugate. This data strongly suggests that the absorption of DON and metabolites by dairy cows is dramatically higher than the <1% reported by Prelusky et al. [107] since urinary excretion is dependent on systemic absorption. It is likely that DON undergoes significant biotransformation to DOM-1 in the rumen prior to being absorbed. Several studies have demonstrated extensive capability of rumen microflora to metabolize DON into the less toxic DOM-1 [37], [51]. It would consequently appear that the rumen affords a high level of protection against DON toxicity.

The pharmacokinetics of DON in sheep have also been investigated. An estimated 6–10% of an oral dose was absorbed systemically. The ratio of DON to conjugated DON in the plasma was approximately 1 to 3, with a very small fraction identified as DOM-1 (predominantly as conjugated form) [109]. These results suggest that the toxin undergoes extensive metabolism when absorbed from the gastrointestinal tract in a first pass since a much higher proportion of the toxin detected in blood occurred as the DON conjugate following oral dosing, as compared to intravenous (*iv)* dosing (73 vs. 20%, respectively). Data from subsequent studies showed, furthermore, that after *iv* administration, 60–85% of the recovered dose occurred as a conjugated metabolite (DON or DOM-1) [110, 111]. This suggests that liver metabolic conjugation of DON is an important step in its elimination from sheep. DON was mainly excreted with urine with an average elimination half-life of 3.2 hours when administered at 5 mgDON/kg b.w. intraruminally. About 50% of the excreted DON was conjugated to glucuronic acid and about 24% was excreted as de-epoxy DON and it's conjugate [110]. Radio-labelled DON was found to be eliminated rapidly from plasma with a half-life of 1.1 hour after *iv* administration of 4 mgDON/kg b.w. [111]. About 0.25% of the dose was excreted into milk and mainly as de-epoxy DON, both conjugated and free. It was interesting to note that following a large single oral or *iv* dose of DON very little DOM-1 (free or conjugated) could be measured in sheep blood [109]. A more prolonged oral exposure (880 mg DON/kg feed for 3 days), however, resulted in DOM-1 being the major component detected (10–20 × DON levels) [111]. The differences in DOM-1 appear to be a result of extended exposure of DON to the rumen microflora, which efficiently metabolized DON to DOM-1. These reports are further evidence of how kinetic and metabolic parameters, and consequently toxicity, can be influenced by the method and duration of administration.

Effect of DON on Ruminants

Field survey data show an association between DON contamination of diets and poor performance in dairy herds. Data from North Carolina from 100 dairy farms representing about 50,000 cows showed that 500-700 μgDON/kg feed was associated with a reduced milk production [112]. It was concluded that DON may simply be a marker mycotoxin for problem feeds.

Exposure to Northern European grass silage containing high amounts of DON caused a toxic syndrome in cattle characterized by an increase in mastitis and laminitis. It is not known to what extent these symptoms are caused by DON exposure only or whether other mycotoxins were also involved [113].

A few feeding trails with DON have been conducted in dairy cattle. CharmLey [25] fed DON contaminated wheat and corn to lactating cows (6 mg/kg and 12 mg/kg of concentrate DM). There was no effect of diet on feed intake or milk production. Cows produced 13% less fat corrected milk. Ingalls [114] performed experiment on lactating dairy cows feeding DON levels up to 14.6 mg/kg concentrate. No effects of DON on feed intake, milk production or milk composition were observed. Trenholm et al. [24] fed non-lactating dairy cows contaminated wheat and DON levels of 1.5 and 6.4 mg/kg were achieved. There was a slight decrease in grain consumption in cows fed the higher DON ration. No effects on animal health were detected. Diaz et al. [115] fed diets, naturally contaminated with 2500 μg/kg DON and 270 μg/kg ZON. Cows consuming diets without a mycotoxin adsorbent had significantly less milk production than cows fed an aluminosilicate mycotoxin adsorbent. This once again demonstrated the possible synergistic effect of combinations of mycotoxins even when present in low levels resulting in impaired performance and health. Researchers in Uruguay [116] studied the effect of DON on udder health. This study demonstrated an increase in somatic cells counts (scc) in cows fed 5 mg/kg of DON in contaminated feed with 333.1 scc x 1000/mL compared to controls with 116.3 scc x 1000/mL. Recent studies in dairy cows demonstrated that feed naturally contaminated with *Fusarium* mycotoxins can affect metabolic parameters and immune function [28, 29]. Total serum protein and globulin levels increased significantly in cows fed the contaminated TMR compared to controls, whereas the albumin:globulin ratio decreased. Serum urea concentrations increased as well as serum sodium and osmolality. Effects on immunity included decreased IgA concentrations, depressed neutrophil phagocytosis and stimulated primary antibody response to ovalbumin immunization.

Beef cattle and sheep appear to tolerate relatively large amounts of DON without obvious deleterious effects. Two trials were conducted at the University of Minnesota with feedlot steers (395 kg) to examine the influence of feed-borne DON on animal performance. Two sources of barley were blended to provide total ration DON levels of 7, 14 and 21 mg/kg [117] and fed to crossbred steers for 144 days. No effects of DON on feed intake or animal performance were observed. DiCostanzo et al. [118] fed crossbred steers (415 kg) contaminated barley for 135 days. DON levels were 0, 6, 12 or 18 mg/kg. No effects of DON on feed intake or animal performance were observed in this trial. Reports from Nebraska indicated similar feed intakes, average daily gains and feed efficiencies compared to controls when sheep (8.5 mg dietary DON/kg feed) or cattle (1 mg dietary DON/kg feed) consumed contaminated diets [119]. Nelson et al. [120] fed feedlot steers and heifers diets containing either 0.2, 2.3 or 10 mgDON/kg for 126 days. The low DON diet was corn based, while the other two contained wheat. Results reported for the low, medium and high DON diets were similar for dry matter intake (DMI), average daily gain (ADG) and feed efficiency. Results for carcass characteristics, serum biochemistry and tissue histology were similar across treatments. Danicke et al., [121] fed *Fusarium* contaminated wheat (10 mg DON and 0.76 mg ZON per kg DM) to growing bulls. The effects on carcass characteristics included reduced dressing percentage, increased weight of the empty gastro-intestinal tract and reduced weight of testicles. Other feeding experiments with beef cattle suggest they can tolerate large concentrations of DON in a feedlot situation without effects on DMI, ADG, or feed to gain ratio [117,122]. Exposing sheep to DON (15.6mg/kg of feed) for 28 days had no effects on average daily gain, hemacytological parameters, or liver function [123].

These reports suggest that beef cattle and sheep may be less sensitive to DON than dairy cattle. Differences could be related to the level of production stress, since mid-lactation, low producing dairy cattle also appear to be more tolerant to DON than high producing dairy cattle in early lactation [124]. Mycotoxin-induced immunosuppression may supplement immunosuppression arising from early lactation, heat and environmental stresses thereby producing more severe effects than would be otherwise expected. Early lactation, high producing cows which experience greater stress, lower immunity, nutrient deficiencies and faster rumen turnover (less mycotoxin degradation in the rumen) may, therefore, be more sensitive to the effects of mycotoxins [124].

Transfer of DON into Milk

Limited or no transfer of DON or metabolites into milk has been reported. Prelusky et al., [107] measured DON in milk following a single oral dose of 920 mg DON to each of 2 lactating cows of similar weight. Free and conjugated DON were present in milk in low amounts (< 4 ng/mL). CharmLey et al., [25] measured DON in milk of mid-lactating cows fed 6 mgDON/kg concentrate (DM) for 10 weeks. No transfer of DON or de-epoxyDON into milk was observed. Concentrations were below the limit of detection (1mg/mL). In a short term study performed by Côté et al. [125], cows were fed corn artificially inoculated with *F. graminearum* and containing 66 mgDON/kg. Unconjugated de-epoxyDON was present in milk at concentrations up to 26 ng/mL. DON was not detected. In recent study by Keese et al., [126] no unmetabolizable DON was detected in milk samples. For cows fed 50% concentrates with 5.3 mgDON/kgDM, DOM-1 concentrations were found in milk at up to 3.2 μg/kg. In the same study the feeding of a high forage low concentrate diet resulted in a higher carry over, suggesting the effect may result from an altered bioavailability of DON from corn silage. The transfer of DON into milk, however, was negligible.

U.S. FDA guidelines for DON levels in feed are 10 mg/kg in grains and by-products destined for chickens or ruminating beef cattle or cattle older than four months provided these products do not exceed 50% of the diet. The guideline is 5 mg/kg in products that will make up no more than 20% of the diet of swine and 5 mg/kg for all other animals in materials that will compose no more than 40% of diet [127].

T-2 toxin

T-2 toxin, a *Fusarium* trichothecene mycotoxin, has been associated with gastroenteritis, intestinal hemorrhages, immunosuppression and death [128-132].

Toxicokinetics of T-2 Toxin

At least 29% of a dose of T-2 toxin is absorbed after administration to dairy cows, but only a small fraction of the parent toxin was recovered [53]. Intravenous administration of T-2 toxin to heifers resulted in a mean elimination phase half-life of 17.4 min and a mean apparent specific volume of distribution of 0.376 l/kg [133]. The fraction of T-2 toxin eliminated as the parent compound in the urine was negligible. When toxic oral doses (up to 3.6 mg/kg feed) were administered, no parent T-2 toxin was detected in plasma or urine.

Effect of T-2 Toxin on Ruminants

Weaver et al. [134] demonstrated that T-2 was associated with feed refusal and gastrointestinal lesions in a cow, but failed to show a hemorrhagic syndrome. Two dairy herds received T-2 contaminated feed from the same supplier [95]. Early lactation cows were more severely affected, showing a lack of appetite, severe and prolonged weight loss, low peak milk production, higher levels of morbidity, and death. In another field case (Jersey herd), with corn produced on the farm, diets contained approximately 350 μgT-2/kg. Cows exhibited diarrhea and a 15% drop in milk production. Bovine infertility and abortions in the final trimester of gestation also have resulted from consumption of feed contaminated with T-2 toxin [21]. Calves consuming 10–50 mgT-2/kg of feed were observed to have ulcers in the abomasum and sloughing of the papilla in the rumen [135]. Dairy cattle fed moldy corn containing 1 mgT-2/kg exhibited hemorrhagic syndrome [131]. A calf intubated with T-2 developed severe depression, hindquarter ataxia, knuckling of the rear feet, listlessness and anorexia [134].

T-2 toxin also affects the immune status of cattle. Mann et al. observed decreases in serum concentrations of IgM, IgG, and IgA, suppression of neutrophil functions and lymphocyte blastogenensis, and in the response of lymphocytes to phytohemagglutinin [136,137]. T-2 toxin was also shown to induce necrosis of lymphoid tissues [138]. Other reports indicated a reduction in white blood cell and neutrophil counts in calves [139].

Carry-Over of T-2 Toxin into Milk

Carry over of T-2 toxin into milk was studied on Holstein cows intubated with 182 mg of T-2 toxin daily for 15 days (corresponding to a concentration in the feed of approximately 50 mg/kg) [140]. Milk samples were taken on the 2nd, 4th, 5th, 8th, 10th, and 12th day of intubation and were analyzed for T-2 toxin by combined gas chromatography-mass spectrometry. All samples except those collected on the 4th and 8th days of intubation contained T-2 toxin in concentrations ranging from 10 to 160 μg/kg (10 to 160 ng/g of milk). The authors suggested that T-2 concentrations found in nature (approximately 2 mg/kg in the feed) would be impossible to detect.

Zearalenone

Zearalenone is a *Fusarium* mycotoxin with a chemical structure similar to estrogen. It produces estrogenic responses in animals and causes infertility, reduced milk production, and hyperestrogenism in cattle [141].

Toxicokinetics of ZON

Ruminal degradation of zearalenone to zearalenol increases molecular polarity affecting absorption and excretion of the toxin [37]. Seeling et al. [142] fed dairy cows *Fusarium* toxin-contaminated wheat (8.21 mg DON/kg dm and 91 μg ZON/kg dm) as well as corn silage (62 μg ZON/kg dm) and grass silage. The ZON metabolites α-ZOL and β-ZOL were recovered at the duodenum in addition to the parent toxin. The recovery of ingested ZON as ZON plus α-ZOL and β-ZOL at the duodenum ranged between 19 and 247%. The portion of ZON (ranging from 29 to 99%) plus α-ZOL and β-ZOL flow at the duodenum increased significantly with increasing ZON intake, whereas the portion of β-ZOL (up to 57%) decreased significantly. In contrast, recoveries in feces of ZON (32-100%), α-ZOL (up to 39%) and β-ZOL (up to 43%) from ZON plus α-ZOL and β-ZOL were independent of ZON intake. It seems that a lower retention time of the feed in the rumen resulting from the increased feed intake may limit the ruminal metabolism of ZON. The relatively steady recovery of ingested ZON as ZON, α-ZOL and β-ZOL in feces at the different levels of ZON intake would suggest a further reduction of ZON by intestinal microorganisms. Furthermore, ZON and its metabolites in the milk were not detectable at daily ZON intake between 75 and

1125 μg /day at DM intake 5.6 and 20.5 kg/day and milk yields (fat corrected milk, FCM) between 10 and 42 kg/day.

In other experiment, *Fusarium* contaminated wheat (8.05 mg and 7.15 mg DON/kg and 0.26 mg and 0.1 mg ZON/kg) was fed to dairy cows [27]. The ZON metabolites α-ZOL and β-ZOL were recovered at the duodenum in addition to the parent toxin. Their recovery as a percentage of ZON-intake ranged from 43% and 132%. The authors concluded that high ZON recovery suggested a lower degradation of the parent toxin in the rumen and/or recovery of some entero-hepatic cycling of biliary ZON/metabolites.

The feeding of 50 or 165 mg ZON daily to three cows for 21 days resulted in the absence of ZON or metabolites in either milk or plasma [143]. A dose of 544.5 mg zearalenone per day given to a single cow for 21 days yielded maximum concentrations of only 2.5 ng ZON/mL and 3.0 ng α-ZOL/mL in the milk. In plasma, up to 3 ng ZON/mL could be detected during the initial 4 days of treatment. At a dose of 1.8 g of zearalenone over a one day feeding period, maximum milk levels of 4.0 ng ZON/mL, 1.5 ng α-ZOL/mL, and 4.1 ng β-ZOL/mL were observed during the initial 2 days. Corresponding maximum levels after a one day dose of 6.0 g ZON were 6.1, 4.0 and 6.6 ng/mL milk on days 2-3. In plasma, peak ZON concentrations (9 and 13 ng/mL at the lower and higher one-day doses, respectively) occurred 12 hr after initial dosing, and declined to negligible levels by days 5-7.

Danicke et al. [121] fed *Fusarium* contaminated wheat (10 mg DON and 0.76 mg ZON/per kg dm) to growing bulls. ZON and its metabolites were not detected in edible tissues.

Effect of ZON on Ruminants

Dairy heifers exposed to ZON have reduced conception rates [144]. A controlled study with non-lactating cows fed up to 500 mg of ZON/day showed no obvious effects except that corpora lutea were smaller in treated cows [145]. In a study with heifers receiving about 13 mg/kg ZON, conception rate was depressed about 25%; otherwise, no obvious effects were noted [146]. In a field study, diets containing about 750 μg/kg ZON and 500 μg/kg DON resulted in poor consumption, depressed milk production, diarrhea, increased reproductive tract infections, and total reproductive failure [147]. In one report from a 150 cow dairy herd, an increased artificial insemination index (decreased fertility) was reported after the herd had been fed moldy hay. Zearalenone concentrations in the hay extracts were reported to be 14 mg/kg. Upon removal of the infected hay, the AI index returned to the previous level. Some researchers have related urinary ZON and it's metabolites to reproductive disorders in sheep and dairy cattle [148-151]. In sheep, zearalenone exposure resulted in lower rates of conception, reduced ovulation, and increased twinning rates. Dairy cow herds with low fertility were found to have higher levels of blood and urinary ZON. Large doses of ZON have been associated with abortions in cattle [152, 153].

It has been suggested that high dietary levels (12 mg/kg of feed) of ZON for extended periods (10 days) may negatively affect reproductive performance of sheep by reducing fertility and ovulation rates [154].

ZON has been shown to suppress chemiluminiscence of bovine neutrophils *in vitro* in a dose-dependant manner [155]. Pretreatment with estrogen receptor blockers, however, did not support the suggestion that ZON affected the cells via estrogen receptor pathways.

Carry Over of ZON into Milk

Transmission of ZON or its metabolites into milk appears to be minimal [143]. The feeding of 50 or 165 mg daily to three cows for 21 days did not result in the presence of ZON or metabolites in milk. A dose of 544.5 mg ZON daily to a single cow for 21 days yielded maximum concentrations of only 2.5 ng ZON/mL and 3.0 ng α-ZOL/mL in the milk. Seeling et al. [142] observed that ZON and metabolites in milk were lower than the detection limits at daily ZON intakes of between 75 mg and 1125 mg for cows yielding between 10 kg and 42 kg fat corrected milk/day.

Fumonisin

Fumonisin B_1 is produced by *F. verticillioides* and *F. proliferatum*. Fumonisin toxicity is thought to result from blockage of sphingolipid biosynthesis resulting in degeneration of nerve tissue.

Toxicokinetics of FB_1

Prelusky et al. [156] observed the rapid disappearance of FB_1 from blood after a 120 min *iv* injection of 0.05 or 0.2 mg FB_1/kg bw. No FB_1 was detected in plasma after oral exposure to 1 or 5 mg/kg bw in the same study. No detectable toxin or conjugates were found in milk [157]. Several researchers have concluded that the carry over rate of fumonisins in milk is approximately 0.05% [158,159]. Systemic absorption based on plasma levels and accumulation of ^{14}C-labelled toxin in tissues has been estimated to be less than 1% of the initial dose [157]. A study conducted with beef cattle fed corn contaminated with a mixture of FB_1 and FB_2 obtained from *F.verticillioides* culture indicated that most of the fumonisins were found unmetabolized in the feces, while only trace quantities were detected in blood and urine [160].

Effect of FB_1 on Ruminants

Liver and kidney injury have been described in cattle exposed to fumonisin with increased serum levels of organ specific diagnostic enzymes, and increased cholesterol and bilirubin levels [161-163].

Dairy cattle (Holsteins and Jerseys) fed diets containing 100 mg/kg fumonisin [164] for 7 days prior to freshening and for 70 days thereafter exhibited lower milk production (6 kg/cow/day), resulting from reduced feed consumption. Increases in serum enzyme activities suggested mild liver injury. Dairy cattle may be more sensitive to fumonisin than are beef cattle because of greater production stress.

Osweiler et al., [165] fed 18 young steers either 15, 31 or 148 mg/kg of fumonisin for 31 days. Mild liver lesions were found in two of six calves exposed to the highest level. Elevated liver enzyme activities indicated liver damage. Lymphocyte blastogenesis was significantly impaired at the end of the feeding period in the group exposed to the highest dose.

Fumonisins at high doses (11.1–45.5 mg/kg bw) have been found to be acutely and fatally nephrotoxic and hepatotoxic in lambs [166]. It should be noted, however, that such experimental levels have not been found in contaminated feeds [55].

It is recommended by the US FDA that human foods should contain no more than 2 to 4 mg of total fumonisins/kg. For dairy cattle, the guideline recommends that contaminated corn or corn-byproducts be limited to no more than 50% of the diet, and that the maximum concentration of total fumonisins in corn and corn by-products are 30 mg/kg for lactating and breeding age cattle and no more that 10 mg/kg for calves [167]. Since the consumption of fumonisin is associated with reduced feed consumption, there is a concern that low levels of fumonisin can interact with other mycotoxins to reduce milk production.

Carry-Over of FB_1 into Milk

Fumonisin carryover from feed to milk is thought to be negligible. Richard et al. [37] fed culture material containing fumonisins (FB_1, FB_2, and FB_3) to 2 mid-lactating Jerseys cows for 14 days at the level of approximately 75 mg/kg. Fumonisins were not detected in any of the milk samples by two separate laboratories using methods with a sensitivity of 5 ng/mL. Except for transient diarrhea at the beginning of the feeding period and an increase in serum cholesterol, clinical and hematologic changes were not observed in the animals. No detectable amounts of fumonisin or conjugates were found in milk after oral or *iv* dosing with FB_1 [157].

Ochratoxin

The ochratoxins are a family of compounds with varying toxicity, and general structure of β-phenylalanine linked by an amide bond to dehydroisocoumarin. The most toxic of the ochratoxins is ochratoxin A. At least seven fungal species of the genus *Aspergillus* and six of the genus *Penicillium* are known to produce ochratoxins. Ochratoxin occurs in wide variety of feedstuffs and is a renal toxin.

Toxicokinetics of OTA

OTA is absorbed passively in non-ionized form which occurs at acid pH [168]. Starch containing diets, therefore, promote OTA absorption more than high fiber diets. These observations also suggest that the forestomachs and the abomasum are the main sites of OTA absorption in the ruminant digestive tract [39].

Blank et al. [169] investigated the metabolism of ochratoxin A feeding various doses (0, 9.5, 19.0, and 28.5 μg OTA/kg bw) to sheep for 29 days. Animals were fed diets consisting of 70% concentrates and 30% grass silage. Significant concentrations of undegraded ochratoxin A were detected in serum of sheep at all levels of OTA tested. Serum concentrations of OTA slightly increased with increasing time of exposure and were linearly dependent on the administered dose of OTA. A constant proportion (6-8%) of the dose was excreted in the urine. The results of this study indicate that even at moderate to low levels of dietary OTA, considerable amounts of the mycotoxin are absorbed by ruminants and may accumulate in tissues.

Hohler et al. [61] indicated that feeding sheep 0, 2 and 5 mg/kg OTA resulted in significant concentrations of OTA and OTA-α in blood serum (10 ng/mL and 3 ng/mL with 2 mg/kg and 80 ng/mL and 15 ng/mL with 5 mg/kg). Total excretion of OTA-α accounted for 70% of OTA intake (9% in feces, 61% in urine), whereas only 1% and 3.8% of OTA intake were found unmetabolized in feces and urine, respectively.

Effect of OTA on Ruminants

The typical signs of ochratoxicosis including renal proximal tubule cell damage and karyomegaly have not been reported in dairy cows [170]. Exposure of cattle to citrinin, a toxin often co-occurring with ochratoxins, however, may cause a haemorrhagic syndrome in cattle [171]. Ochratoxins did not induce any significant clinical symptoms in cattle when barley naturally-contaminated with OTA (390–540 µg/kg) and low levels of AFB_1 (12–13 µg/kg) were fed to12-week-old calves [172]. OTA induced apoptosis of bovine lymphocytes *in vitro* [173].

Lambs dosed with 1mg of OTA/kg died after 24 hrs [174]. Pulmonary edema, friable livers and foetus lesions were observed. Goats were dosed daily with 1, 2 or 3 mg/kg feed OTA. Goats administered the high dose died after 6 days of exposure while these receiving 1 and 2 mg OTA/kg showed no obvious sign of toxicity for the duration of the 14 day experiment.

Carry-over of OTA into Milk

A survey on the presence of OTA in raw bulk milk was conducted in 2003 in the northwest of France in 132 randomLy selected farms [105]. OTA was detected in 3 out of 264 samples at the levels of 5 to 8 ng/L. In a Swedish survey, 36 milk samples were analyzed and OTA was found in 5 (14%) samples with a range of 10-40 ng/mL [175]. Scaug [176] detected OTA above 10 ng/l in 6 out of 40 samples form traditional Norwegian farms and 5 out of 47 from organic farms with maximum concentrations of 28 ng/l and 58 ng/l for organic and traditional farms respectively. A survey of 50 organic and traditional farms in the UK found no detectable OTA [177].

Ergot Alkaloids

Ergot alkaloids are produced by *Claviceps* spp. which can infest cereal crops and grasses. Animals consume ergot by eating the sclerotia present in contaminated feed. Ergovaline is measured as a marker toxin for the presence of ergot alkaloids [17]. Ergovaline acts as a dopamine receptor agonist, and this principle mechanism explains most of the clinical symptoms associated with fescue toxicosis, such as reduced milk production, low conception rates, and loss of bodyweight [178]. Another typical symptom is intolerance to high environmental temperatures, causing a pathological hyperthermia and lipid necrosis during the summer season. This hyperthermia is thought to result from peripheral vasoconstriction, a common symptom of ergot alkaloid toxicity. Laminitis and foot gangrene ('fescue foot') also suggest the presence of more than one alkaloid in infected tall fescue.

Toxicokinetics of Ergot Alkaloids

Specific kinetic data for individual ergot alkaloids in food producing animals are not available. Some data exists for ergovaline in horses [179]. It would appear that the rumen does not protect against alkaloid toxicity [67,180]. It is generally assumed that the hydrophilic amides, such as ergometrine, are rapidly absorbed whereas the less water-soluble alkaloids of the ergotamine group have a lower oral bioavailability which has been estimated to be 62 % in humans [181]. Whittemore et al. [182,183], however, described oral availability of approximately 90 % for unspecified ergot alkaloids in pigs. Ergot alkaloids undergo extensive first-pass metabolism in the liver following oral ingestion. A prominent example is methysergide, which has an oral bioavailability of approximately 13% and is converted in the liver into the active methylergometrine. Ergot alkaloids are biotransformed in the liver, but the structures of many metabolites remain to be elucidated although it is likely that they retain biological activity [184,185]. Excretion is often biphasic, depending on the degree of enterohepatic recirculation. Investigations in rabbits with dihydroergocryptine, dihydroergotamine and dihydroergocornine given intravenously at a dosage of 20 mg/kg b.w. showed that the highest concentrations were measurable in liver, kidney, spleen and muscle tissue. None of these dihydro-compounds were able to cross the blood-brain barrier. Total body elimination was calculated to be < 20 hours [186].

Effect of Ergot Alkaloids on Ruminants

All domestic animals are susceptible to ergot alkaloid toxicosis. Cattle seem to be the most susceptible. Two well recognized types of ergotism have been reported in animals including acute toxicity characterized by convulsions and chronic toxicity characterized by gangrene. Other symptoms of ergotism in cattle include hyperthermia and agalactia. The symptoms manifested depend on the type of ergot consumed and the ratios of the major toxic alkaloids present in the ergot. The responses of animals to feed-borne ergot are usually quite variable and depend on variations in alkaloid content, frequency of ingestion, quantity ingested, climatic conditions preharvest and the species of ergot involved [180].

Lameness, sometimes leading to gangrene, is a common symptom observed in cattle when the feed contains more than 10 g ergot/kg. The symptoms are more pronounced when the animals are kept outside under cold weather conditions [187]. Hyperthermia (summer fescue toxicosis) in cattle has been characterized by symptoms of weight loss, excessive salivation and heat stress [188]. Fat necrosis in cattle is a condition in which areas of the fat are hardened resulting in constriction of internal organs, reduced serum cholesterol, and elevated serum amylase [189].

Four out of 6 animals orally administered ergotamine tartrate (1 mg/kg bw daily) died within 10 days [190]. The animals became acutely ill within 1 - 2 days with the principal signs being anorexia, hyperventilation, cold extremities, excessive salivation, and occasionally tongue necrosis. Post-mortem examination of the most seriously affected animals revealed extensive intestinal inflammation. In an Australian study [191], cattle (heifers and young steers) were fed grain sorghum for 28 days. Analysis indicated 0.44% of ergot sclerotia with an average alkaloid content of the primary grain heads of 3.5 mg/kg d.m. and of tiller heads of 4.6 mg/kg d.m. The alkaloid concentration of leaf and stalks was < 1 mg/kg dry matter. The estimated daily intake was about 35 mg (total alkaloids) per animal (the body weight varied between 400 - 405 kg for the heifers and 283- 295 kg for the young steers). The major alkaloid detected was dihydroergosine, which is consistent with reported alkaloid production by *C. africana*. None of the animals showed clinical signs of intoxication but serum prolactin levels were significantly decreased, particularly in the young steers, which already have lower physiological levels than heifers. An outbreak of bovine abortion associated with the ingestion of ergot was reported by Appleyard [192]. Eleven out of 36 cows in late pregnancy aborted in 7 - 11 days following introduction to a rye grass pasture heavily infested with ergot. At least 25 % of the rye seed heads contained sclerotia of *C. purpurea,* with up to 8 sclerotia present on one seed head. The individual sclerotia contained

1.57 mg total ergot alkaloids/g which included of 67 % ergotamine, 17 % ergotaminine and smaller amounts of ergometrine. Ergocryptine and ergosine were detectable in these sclerotia as a mixture of their natural occurring isomers. Cattle consuming tall fescue contaminated with endophytic fungi such as *A. lolii* have shown symptoms of staggers, excitability, increased rectal temperature, increased respiration rate, and loss of body weight [193].

Four lambs were administered aqueous suspensions of milled ergot (from *C. purpurea)* through a stomach tube, over a 2-month period [194]. The doses ranged from 0.12 to 0.75 g sclerotia/kg b.w. The sclerotia contained approximately 4 g ergolines/kg including ergotamine (15%), ergosine (35%), and ergocristine (5% each). One lamb dosed with 0.12 g sclerotia/kg b.w. was kept indoors at 15-17 °C and did not develop any symptoms. The other 3 animals, given higher doses and kept outdoors became ill after 2-6 days with signs that included dullness, inappetence, high pulse rate, diarrhoea, edema of the hind legs and tail, and lameness. Post-mortem findings included inflammation and necrosis of the forestomach and intestinal mucosa.

Carry-Over of Ergot Alkaloids into Milk

The carry-over following exposure of dairy cows to 50 g ergot sclerotia per animal (approximately 400 kg b.w.) resulted in milk concentrations of total alkaloids reaching 0.086 mg/L [195]. Wolff et al. [196] fed dairy cows an alkaloid concentration corresponding to 1835 μg/animal/day (equivalent to approximately 3 μg/kg bw) and failed to detect residues of ergot alkaloids in milk. The authors concluded that the carry-over rate into milk is less than 10 % of the applied dose.

CONCLUSION

There are multiple factors contributing to adverse effects arising from feed-borne mycotoxins in ruminants including feeding management, farm management, health issues and stress factors. The complexity and wide range of ingredients in ruminant diets can result in combinations of mycotoxins with synergistic effects. A number of mycotoxins are inactivated by the rumen microflora and transformed to less toxic metabolites. Some mycotoxins can pass unchanged while others can be converted into metabolites that retain biological activity or have even higher toxicity. Bioavailability of natural occurring mycotoxins to rumen microflora can be influenced by the feed matrix, by conjugation and by impaired rumen function caused by metabolic disorders. These factors will increase the time required for degradation and subsequently increase chance of the mycotoxin escaping rumen metabolism and becoming available to the animal.

The degree of feed contamination can also be underestimated because of conjugated mycotoxins. The presence of these compounds can escape routine detection methods but can be released after hydrolysis in the digestive tract.

Rumen function can be negatively affected by the presence of mycotoxins. This can be characterized by decreased cellulose degradation, slowed rumen motility, altered volatile fatty acid production, ammonia production and microbial protein synthesis. The capacity of the rumen to inactivate mycotoxins can be influenced by the type of diet, the use of feed additives and medications, microbial composition and population, rumen passage rate and some stress factors such as decreased feed intake in transition cows.

The rate of transfer of mycotoxins from blood into milk might be affected by level of milk production as well as by systemic or local disease resulting in impaired transport barrier. The effects of mycotoxins on immunity can result in lower resistance to local or systemic disorders which are normally difficult to estimate under commercial farm conditions.

Symptoms of mycotoxicosis in a dairy herd may be non-specific and wide ranging, and vary depending on the mycotoxins involved and their interaction with other stress factors. The largest economic losses come from impaired performance and reproduction. Such subclinical effects may not get noticed at the farm level. Efforts should be taken towards good feed management to reduce diet contamination as well as increasing cow immunity and rumen health to help combat decreased disease resistance arising from feed-borne mycotoxins.

References

[1] Coulombe, R.A. Biological Action of Mycotoxins. *Journal of Dairy Science*, 1993, 76, 880-891.

[2] Whitlow, L.W.; Hagler W.M.Jr. Mold and mycotoxin issues in dairy cattle: effects, prevention and treatment. North Carolina State University, 2008. Available at URL: http://www.cals.ncsu.edu/an_sci/extension.

[3] CAST. *Mycotoxins: Risks in plant, animal, and human systems*. Task Force Report No. 139. Council for Agricultural Science and Technology, Ames, IA, 2003.

[4] Cole R.J.; Kirksey J.W.; Dorner J.W.; Wilson D.M.; Johnson JrJ.; Bedell D.; Springer J.P.; Chexal K.K.; Clardy J.; Cox R.H. Mycotoxins produced by *Aspergillus fumigatus* isolated from silage. *Annual Nutrition Alimentation*, 1977, 31, 685–691.

[5] Schneweis I.; Meyer K.; Hormansdorfer S.; Bauer J. Metabolites of *Monascus ruber* in silages. *Journal of Animal Physiology and Animal Nutrition* (Berl), 2001, 85, 38–44.

[6] Escoula L. Toxinogenic moulds of silage. IV. Patulin production in liquid medium using fungus species isolated in silages *Annales de recherchers veterinaries,* 1975, 6, 303–310.

[7] Fink-Gremmels, J. Mycotoxins in cattle feeds and carry-over to dairy milk: A review. *Food Additives and Contaminants*, 2008a, 25, 172-180.

[8] Garon D.; Richard E.; Sage L.; Bouchart V.; Pottier D.; Lebailly P. Mycoflora and multimycotoxin detection in corn silage: experimental study. *Journal of Agricultural and Food Chemistry*, 2006, 54, 3479–3484.

[9] O'Brien M.; Nielsen K.F.; O'Kiely P.; Forristal P.D.; Fuller H.T.; Frisvad J.C. Mycotoxins and other secondary metabolites produced *in vitro* by *Penicillium paneum* Frisvad and *Penicillium roqueforti* Thom isolated from baled grass silage in Ireland. *Journal of Agricultural and Food Chemistry*, 2006, 54, 9268–9276.

[10] Whitlow, L.W.; Hagler W.M.Jr.; Hopkins B.A. Mycotoxin occurrence in farmer submitted samples of North Carolina feedstuffs: 1989-1997. *Journal of Dairy Science*, 1998, 81(Abstr.),1189.

[11] Driehuis F.; Spanjer M. C.; Scholten J.M.; te Giffel M.C. Occurrence of mycotoxins in feedstuffs of dairy cows and estimation of total dietary intakes. *Journal of Dairy Science,* 2008, 91, 4261–4271.

[12] Auerbach, H.; Oldenburg E.; Weissbach F. Incidence of *Penicillium roqueforti* and roquefortine C in silages. *Journal of the Science of Food and Agriculture*, 1998b, 76, 565–572.

[13] Schneweis, I.; Meyer K.; Hörmansdorfer S.; Bauer J. Mycophenolic acid in silage. *Applied and Environmental Microbiology*, 2000, 66, 3639– 3641.

[14] Cheeke P.R. Endogenous toxins and mycotoxins in forage grasses and their effects on livestock. *Journal of Animal Science*, 1995, 73, 909–918.

[15] Tor-Agbidye, J.; Blythe L.L.; Craig A.M. Correlation of endophyte toxins (ergovaline and lolitrem B) with clinical disease: fescue foot and perennial ryegrass staggers. *Veterinary and Human Toxicology*, 2001, 43, 140–146.

[16] Fink-Gremmels, J. Mycotoxins in forages. In: Diaz, D.E. (Ed.), *The Mycotoxin Blue Book.* Nottingham University Press, Nottingham, United Kingdom, 2005, 249–268.

[17] Botha, C.J.; Naude T.W.; Moroe M.L.; Rottinghaus G.E. Gangrenous ergotism in cattle grazing fescue (*Festuca elatior L.*) in South Africa. *Journal of the South African Veterinary Association*, 2004, 75, 45–48.

[18] Nawaz S.; Scudamore K.A.; Rainbird S.C. Mycotoxins in ingredients of animal feeding stuffs: I. Determination of Alternaria mycotoxins in oilseed rape meal and sunflower seed meal. *Food Additives and Contaminants,* 1997, 14, 249–262.

[19] Scudamore KA.; Nawaz S.; Hetmanski M.T. Mycotoxins in ingredients of animal feeding stuffs: II. Determination of mycotoxins in maize and maize products. *Food Additives and Contaminants,* 1998a, 15, 30–55.

[20] Scudamore K.A.; Nawaz S.; Hetmanski M.T.; Rainbird S.C. Mycotoxins in ingredients of animal feeding stuffs: III. Determination of mycotoxins in rice bran. *Food Additives and Contaminants*, 1998b, 15, 185–194.

[21] Placinta C.M.; D'Mello J.P.F.; MacDonald E.K. A review of worldwide contamination of animal feed with *Fusarium* mycotoxins. *Animal Feed Science and Technology*, 1999, 78, 21–37.

[22] Harvey, R.B.; Kubena L.F.; Huff W.E; Elissalde M.N.; Philips T.D. Hematologic and immunologic toxicity of deoxynivalenol contaminated diets to growing chickens. *Bulletin of Environmental Contamination and Toxicology,* 1991, 40, 410–416.

[23] Trenholm, H. L.; Prelusky D.B. Reducing mycotoxins in animal feeds. Publ. #1827E. *Agriculture Canada Publication,* Ottawa, Ontario, Canada, 1988.

[24] Trenholm, H. L.; Thompson B.K.; Hartin K.E.; Greenhalgh R.; McAlister A.J. Ingestion of vomitoxin (deoxynivalenol) contaminated wheat by nonlactation dairy cows. *Journal of Dairy Science*, 1985, 68, 1000–1005.

[25] CharmLey E.; Trenholm H.L.; Thompson B.K.; Vudathala D.; Nicholson J.W.G.; Prelusky D.B.; CharmLey L.L. Influence of level of deoxynivalenol in the diet of dairy cows on feed intake, milk production, and its composition. *Journal of Dairy Science,* 1993, 76, 3580– 3587.

[26] Whitlow, L. W.; Nebel R. L.; Hagler W.M., Jr. The association of deoxynivalenol in grain with milk production loss in dairy cows. In: Llewellyn G.C.; Dashek W.V.; O'Rear C.E., ed. *Biodeterioration Research 4, Mycotoxins, Wood Decay, Plant Stress,*

Biocorrosion and General Biodeterioration. Plenum Press, New York, NY, 1994, 131–140.

[27] Danicke, S.; Matthaus K.; Lebzien P.; Valenta H.; Stemme K.; Ueberschar K-H.; Razzazi-Fazeli E.; Bohm J.; Flachowsky G. Effects of *Fusarium* toxin-contaminated wheat grain on nutrient turnover, microbial protein synthesis and metabolism of deoxynivalenol and zearalenone in the rumen of dairy cows. *Journal of Animal Physiology and Animal Nutrition* (Berl.), 2005, 89, 303–315.

[28] Korosteleva, S.N.; Smith T.K.; Boermans H.J. Effects of feedborne *Fusarium* mycotoxins on performance, metabolism and immunity of dairy cows. *Journal of Dairy Science,* 2007, 90, 3867–3873.

[29] Korosteleva, S.N.; Smith T.K.; Boermans H.J. Effects of feed naturally contaminated with *Fusarium* mycotoxins on metabolism and immunity of dairy cows. *Journal of Dairy Science*, 2009, 92, 1585–1593.

[30] Berthiller, F.; Asta C.D.; Schuhmacher R.; Lemmens M.; Adam G.; Krska R. Masked mycotoxins: Determination of a deoxynivalenol glucoside in artificially and naturally contaminated wheat by liquid chromatography-tandem mass spectrometry. *Journal of Agricultural and Food Chemistry,* 2005, 53, 3421−3425.

[31] Zhou, B.; Li T.; Gillespie J.; He G-Q.; Horsley R.; Schwarz P. Doehlert matrix design for optimization of the determination of bound deoxynivalenol in barley grain with trifluoroacetic acid (TFA). *Journal of Agricultural and Food Chemistry*, 2007, 55, 10141–10149.

[32] Sasanya J.J.; Hall C.; Wolf-Hall C. 2008. Analyses of deoxynivalenol, masked deoxynivalenol, and *Fusarium graminearum* pigment in wheat samples, using liquid chromatography-UV-mass spectrometry. *Journal of Food Protection*, 2008, 71, 1205-13.

[33] Gareis, M.; Bauer J.; Thiem J.; Plank G.; Gedek B. Cleavage of zearalenone-glycoside, a "masked" mycotoxin, during digestion in swine. *Zentralbl Veterinarmed*, 1990, 37, 236–240.

[34] Hult, K.; Teiling A,; Gatenback S. Degradation of ochratoxin A by the ruminant. *Applied and Environmental Microbiology*, 1976, 32, 443–444.

[35] Ribelin, W.E.; Fukushima K.; Still P.E. The toxicity of ochratoxins to ruminants. *Canadian Journal of Comparative Medicine*, 1978, 42, 172– 177.

[36] Bodine, A.B.; Mertens D.R. Toxicology, metabolism, and physiological effect of aflatoxin in the bovine. In Diener, U.L., Asquith, R.L. and Dickens, J.W. Eds. *Aflatoxin and Aspergillus flavus in Corn*. Alabama, Alabama Ag. Exp. Sta., Auburn University, 1983, 46-50.

[37] Kiessling, K.; Pattersson H.; Sandholm K.; Olsen M. Metabolism of aflatoxin, zearalenone, and three trichothecenes by intact rumen fluid, rumen protozoa. *Applied and Environmental Microbiology*, 1984, 47, 1070–1073.

[38] Wogan G.N. Chemical nature and biological effects of the aflatoxins. *Bacteriological Reviewers*, 1966, 2, 460-470.

[39] Jouany, J.; Diaz, D. Effects of mycotoxins in ruminants. In: Diaz, D.E. (Ed.), *The Mycotoxin Blue Book*. Nottingham University Press, Nottingham, United Kingdom, 2005, 295–321.

[40] Engel V.G. and H Hagemeister. Untersuchungen uber den verbleib von aflatoxin B_1 im verdauungstrakt von kunen. *Milchwissenenschaft.,* 1978, 33, 21-23.

[41] Mathur C.F.; Smith R.C.; Hawkins G.E. Growth and morphology of *Streptococcus bovis* and of mixed rumen bacteria in the presence of aflatoxin B_1, *in vitro*. *Journal of Dairy Science*, 1976, 59, 455-458.

[42] Westlake, K., Mackie R.I.; Dutton M.F. *In vitro* metabolism of mycotoxins by bacterial, protozoal and ovine ruminal fluid preparations. *Animal Feed Science and Technology*, 1989, 25, 169-178.

[43] Trucksess, M.W.; Richard J.L.; Stoloff L.; McDonald J.S.; BrumLey W.C. Absorption and distribution patterns of aflatoxicol and aflatoxin B_1 and M_1 in blood and milk of cows given aflatoxin B_1. *American Journal of Veterinary Research*, 1983, 44, 1753-1756.

[44] Auerbach, H.; Maas R.F.M.; Op Den Camp H.J.M.; Pol A.; Fink-Gemmels J. Biodegradation of aflatoxin B_1 by bovine rumen microorganisms *in vitro* and its effects on rumen fermentation. *Revue de Medecine Veterinaire,* 1998a, 149, 573.

[45] Upadhaya S.D.; Sung H.G.; Lee C.H.; Lee S.Y.; Kim S.W.; Cho K.J.; Ha J.K. Comparative study on the aflatoxin B_1 degradation ability of rumen fluid from Holstein steers and Korean native goats. *Journal of Veterinary Science*, 2009, 10, 29-34.

[46] Hagler Jr., W.M.; Mirocha C.J.; Pathre S.V.; Behrens J.C. Identfication of the naturally occurring isomer of zearalenol produced by *Fusarium roseum* "gibbosum" in rice culture. *Applied and Environmental Microbiology*, 1979, 37, 849-853.

[47] Tiemann, U.; Viergutz T.; Jonas L.; Schneider F. Influence of the mycotoxins a- and β-zearalenol and deoxynivalenol on the cell cycle of cultured porcine endometrial cells. *Reproductive Toxicology*, 2003, 5506, 1-10.

[48] Swanson, S.P.; Nicoletti J.; Rood H.D., Jr.; Buck W.; Cote L.M. Metabolism of three trichothecene mycotoxins, T-2 toxin, diacetoxyscirpenol, and deoxynivalenol, by bovine rumen microorganisms. *Journal of Chromatography*, 1987a, 414, 335 342.

[49] Swanson, S.P.; Rood H.D.; Behrens J.C.; Sanders P.E. Preparation and characterization of the deepoxy trichothecenes: deepoxy HT-2, deepoxy T-2 triol, deepoxy T-2 tetraol, deepoxy 15-monoacetoxyscirpenol, deepoxy scirpenol. *Applied and Environmental Microbiology,*1987b, 53, 2821-2826.

[50] Côté, L.M.; Nicoletti J.; Swanson S.P; Buck W.B. Production of deepoxy-deoxynivalenol (DOM-1), a metabolite of deoxynivalenol, by *in vitro* rumen incubation. *Journal of Agricultural and Food Chemistry*, 1986b, 34, 458-460.

[51] King, R.R.; McQueen R.E; Levesque D.; Greenhalgh R. Transformation of deoxynivalenol (vomitoxin) by rumen microorganisms. *Journal of Agricultural and Food Chemistry,* 1984, 32, 1181–1183.

[52] Sato, N.; Ueno Y. *Mycotoxins in human and animal health.* Pathotox Publishers, Park Forest South, IL,USA, 1977.

[53] Yoshizawa, T.; Mirocha C.J.; Behrens J.C.; Swanson S.P. Metabolic fate of T-2 toxin in a lactating cow. *Food and Cosmetic Toxicology,* 1981, 19, 31–39.

[54] Yoshizawa, T.C.; Okamoto K.; Sakamoto T.; Kuwamura K. *In vivo* metabolism of T-2 toxin; a trichothecene mycotoxin. On the formation of deepoxydation products. . *Japanese Association of Mycotoxicologists*, 1985, 21, 9-12.

[55] Hussein, H.S.; Brasel J.M. Toxicity, metabolism, and impact of mycotoxins on humans and animals. *Toxicology*, 2001, 167, 101–134.

[56] Caloni, F.; Spotti M.; Auerbach H; Op den Camp H; Grimmels J.F.; Pompa G. *In vitro* metabolism of fumonisin B_1 by ruminal microflora. *Veterinary Research Communications*, 2000, 24, 379–387.

[57] Gurung N.K.; Rankins D.L.,Jr.; Shelby R.A. *In vitro* ruminal disappearance of fumonisin B_1 and its effects on *in vitro* dry matter disappearance. *Veterinary and Human Toxicology*, 1999, 41,196-199.

[58] Follmann, W.; Hillebrand I.E.;, Creppy E.E.; Bolt H.M. Sister chromatid exchange frequency in culture isolated porcine urinary bladder epithelial cells (PUBEC) treated with ochratoxin A and *a*. *Archives of Toxicology*, 1995, 69, 280-286.

[59] Galtier P.; Alvinerie M. *In vitro* transformation of ochratoxin A by animal microbial floras. *Annales de recherchers veterinaries*, 1976, 7, 91-98.

[60] Muller K. Einfluss der futtering und anderer faktoren auf den umsatz von ochratoxin A in pansenflussigkeit *in vitro* und *in vivo*. PhD. Thesis, Hohenheim University, Stuttgart, Germany, 1995.

[61] Hohler, D.; Sudekum K.H.; Wolffram S.; Frohlich A.A.; Marquardt R.R. Metabolism and excretion of ochratoxin A fed to sheep. *Journal of Animal Science,* 1999, 77, 1217-1223.

[62] Xiao, H.; Marquardt R.R.; Frohlich A.A.; Phillips G.D.; Vitti T.G. Effect of a hay and a grain diet on the rate of hydrolysis of ochratoxin A in the rumen of sheep. *Journal of Animal Science*, 1991a, 69, 3706-3714.

[63] Xiao, H.; Marquardt R.R.; Frohlich A.A.; Phillips G.D.; Vitti T.G. Effect of a hay and a grain diet on the bioavailability of ochratoxin A in the rumen of sheep. *Journal of Animal Science*, 1991b, 69, 3715-3723.

[64] Muller, H.M.; Lerch C.; Muller K.; Eggert W. Kinetic profiles of ochratoxin A and ochratoxin *a* during *in vitro* incubation in buffered forestomach and abomasal contents from cows. *Journal of Natural Toxins,* 1998, 6, 251-258.

[65] Muller, H.M.; Muller K.; Steingass H. Effect of feeding regime on the metabolism of ochratoxin A during the *in vitro* incubation in buffered rumen fluid from cows. *Arch.Tierernahr.*, 2001. 54, 265-279.

[66] Ozpinar H.; Bilal T.; Abas I.; Kutay C. Umsatz von ochratoxin A in pansenflussigkeit *in vitro*. Facta Universitas. *Series Medicine and Biology,* 2002, 9, 66-69.

[67] Fink-Gremmels, J. The role of mycotoxins in the health and performance of dairy cows. *Veterinary Journal*, 2008b, 176, 84–92.

[68] Cook, W.O.; Richard J.L.; Osweiller G.D.; Trampel D.W. 1986. Clinical and pathologic changes in acute bovine aflatoxicosis: rumen motility and tissue and fluid concentrations of aflatoxins B_1 and M_1. *American Journal of Veterinary Research*, 1986, 47, 1817–1825.

[69] Mertens, D.R. Biological effects of mycotoxins upon rumen fermentation and lactating dairy cows. In: *Interactions of mycotoxins in animal production*. National Academy Press, Washington, DC, 1979, 118-136.

[70] Helferich, W.G.; Garrett W.N.; Hsieh D.P.H.; Baldwin R.L. 1986. Feedlot performance and tissue residues of cattle consuming diets containing aflatoxins. *Journal of Animal Science*, 1986, 62, 691–696.

[71] Diekman, D.A.; Green M.L. Mycotoxins and reproduction in domestic livestock. *Journal of Animal Science,* 1992, 70, 1615-1627.

[72] Danicke, S. Effects of *Fusarim* toxin contaminated wheat grain and of a detoxifying agent on rumen physiological parameters and *in sacco* dry matter degradation of wheat straw and Lucerne hay in wethers. *Journal of Animal and Feed Sciences*, 2002, 11, 437-451.

[73] Mackie, R.I.; White B.A. Recent advances in rumen microbial ecology and metabolism: Potential impact on nutrient output. *Journal of Dairy Science,* 1990, 73, 2971-2995.

[74] May H.D.; Wu Q.; Blake C.K. Effects of the *Fusarium* spp. mycotoxins fusaric acid and deoxynivalenol on the growth of *Ruminococcus albus* and *Methanobrevibacter ruminantium*. *Canadian Journal of Microbiology,* 2000, 46, 692-699.

[75] Westlake, K.; Mackie R.I.; Dutton M.F. Effects of several mycotoxins on specific growth rate of *Butyrivibro fibriosolvens* and toxin degradation *in vitro*. *Applied and Environmental Microbiology,* 1987, 53, 613-614.

[76] Escoula L. Moisissures toxicogenes des fourrages ensiles. 1.Presence de patuline dans le fronts de coupe d'ensilages. *Annales de recherchers veterinaries,* 1974, 5, 423-432.

[77] Wooters, M.F.A.; Speijers C.J.A. Patulin. *WHO Food Additives series IPCS INCHEM,* 2003, 35, 1-22.

[78] Madhyastha S.; Maraquadt R.R.; Mast A; Borsa J.; Frohlich A.A. Comparison of toxicity of different mycotoxins to several species of bacteria and yeast: Use of *Bacillus brevis* in a disc diffusion assay. *Journal of Food Protection*, 1994, 57, 48-53.

[79] Escoula L. Patulin production by *Penicillium granulatum* and inhibition of rumen flora. *Journal of Emvironmental Pathology, Toxicology and Oncology*, 1992, 11, 45-48.

[80] Tapia M.O.; Stern M.D.; Koski R.L.; Bach A.; Murphy M.J. Effects of patulin on rumen microbial fermentation in continuous culture fermenters. *Animal Feed Science and Technology*, 2002, 97, 239-246.

[81] Sabbioni, G.; Sepai O. Determination of human exposure to aflatoxins. In Sinha K.K.; Bhatnagar D. ed. *Mycotoxins in Agriculture and Food Safety*. Marcel Dekker Inc., New York, 1998, 183–226.

[82] Kuilman M.E.; Maas R.F.; Fink-Gremmels J. Cytochrome P450-mediated metabolism and cytotoxicity of aflatoxin B_1 in bovine hepatocytes. *Toxicology in Vitro,* 2000, 14, 321–327.

[83] Wood, G.E. Aflatoxin M_1. In: Sharma R.P.; Salunke D.K., ed. *Mycotoxins and Phypoalexins*. CRC Press Inc, Boca Raton, Fl, USA, 1991, 145-164.

[84] Cockcroft, P.D. Sudden death in dairy cattle with putative acute aflatoxin B_1 poisoning. *Veterinary Record*, 1995, 136-248.

[85] Colvin, B.M.; Harrison L.R.; Gosser H.S.; Hall R.F. 1984. Aflatoxicosis in feeder cattle. *Journal of the American Veterinary Medical Association*, 1984, 184, 956–958.

[86] Nibbelink, S.K. Aflatoxicosis in food animals: A clinical review. *Iowa State Univ.Vet.,*1986, 48, 28-31.

[87] Choudhary, P.L.; Sharma R.S.; Borkhataria V.N.; Desai M.C. Effect of feeding aflatoxin B_1 on feed consumption through naturally contaminated feeds. *Indian Journal of Animal Science*, 1998, 68, 400–401.

[88] Applebaum, R.S.; Brackett R.E; Wiseman D.W.; Marth E.L. Responses of dairy cows to dietary aflatoxin: feed intake and yield, toxin content, and quality of milk of cows treated with pure and impure aflatoxin. *Journal of Dairy Science* 1982, 65, 1503-1508.

[89] [Masri, M.S.; Garcia V.C.; Page J.R. The aflatoxin M_1 content of milk from cows fed known amounts of aflatoxin. *Veterinary Research*, 1969, 84, 146-147.

[90] Patterson, D.S.P.; Anderson P.H. Recent aflatoxin feeding experiments in cattle. *Veterinary Record,*1982, 110, 60-61.

[91] Guthrie, L.D. Effects of Aflatoxin in corn on production and reproduction in dairy cattle. *Journal of Dairy Science*, 1979, 62 (abstr.), 134.

[92] Paul, P.S.; Johnson D.W.; Mirocha C.J.; Soper F.F.; Thoen C.C.; Muscoplat C.C.; Weber A.F. *In vitro* stimulation of bovine peripheral blood lymphocytes: suppression of phytomitogen and specific antigen lymphocyte responses by aflatoxin. *American Journal of Veterinary Research*, 1977, 38, 2033–2035.

[93] Bodine, A.B.; Fisher S.F.; Gangjee S. Effect of aflatoxin B_1 and major metabolites on phytohemeagglutininstimulated lymphoblastogenesis of bovine lymphocytes. *Journal of Dairy Science,* 1984, 67, 110–114.

[94] Garrett, W.N.; Heitman H., Jr.; Booth A.N. Aflatoxin toxicity in beef cattle. *Experimental Biology and Medicine*, 1968, 127, 188-190.

[95] Whitlow L.W.; Hagler W.M.Jr. Mycotoxins: A Review of Dairy Concerns *Mid-South Ruminant Nutrition Conference*, 2005.

[96] Neathery, M.W.; Moors W.H.; Wyatt R.D.; Miller W.I.; Gentry R.P.; George L.W. Effects of dietary aflatoxin on performance and zinc metabolism in dairy calves. *Journal of Dairy Science*, 1980, 63,789-799.

[97] Harvey, R.B.; Edrington T.S.; Kubena L.F.; Elissalde M.H.C.D.E; Rottinghaus G.E. Effect of aflatoxin and diacetoxyscirpenol in ewe lambs. *Bulletin of Environmental Contamination and Toxicology*, 1995, 54, 325–330.

[98] Fernandez, A.; Belio R.; Ramos J.J; Sanz M.C.; Saez T. Aflatoxins and their metabolites in the tissues, feces and urine from lambs feeding on an aflatoxin-contaminated diet *Journal of the Science of Food and Agriculture*, 1997, 74, 161–168.

[99] Ramos, J.J.; Fernandez A.; Saez T.; Sanz M.C.; Marca M.C. Effect of aflatoxicosis on blood mineral constituents of growing lambs. *Small Ruminant Research*, 1996, 21, 233–238.

[100] Fernandez, A. ; Ramos J.J. ; Saez T. ; Verde M.T. Changes in the coagulation profile of lambs intoxicated with aflatoxin in their feed. *Veterinary Research*, 1995, 26, 180–184.

[101] Fernandez, A.; Hernandez M.; Verde M.T.; Sanz M. Effect of aflatoxin on performance, hematology, and clinical immunology in lambs. *Canadian Journal of Veterinary Research*, 2000, 64, 53–58.

[102] Van Egmond, H.P. Mycotoxins in dairy products. *Elsevier Science Pub.* Co., Ltd. New York, 1989.

[103] EFSA. Opinion of the Scientific Panel on Contaminants in the Food Chain on a request from the Commission related to Aflatoxin B_1 as undesirable substance in animal feed. *European Food Safety Authority* (EFSA), 2004a, EFSA-Q-2003-035.

[104] Veldman A,; Meijst J.A.C.; Borggreve G.J.; Heeres-van Tol J.J. Carry-over of aflatoxin from cow's food to milk. *Animal Production*, 1992, 55, 163–168.

[105] Boudra H.; Barnouin J. ; Dragacci S. ; Morgavi D.P. Aflatoxin M_1 and ochratoxin A in raw bulk milk from French dairy herds. *Journal of Dairy Science*, 2007, 90, 3197-201.

[106] Battacone G.; Nudda A.; Cannas A.; Cappio Borlino A.; Bomboi G.; Pulina G. Excretion of aflatoxin M_1 in milk of dairy ewes treated with different doses of aflatoxin B_1. *Journal of Dairy Science,* 2003, 86, 2667–2675.

[107] Prelusky, D. B.; Trenholm H.L.; Lawrence G.A.; Scott P.M. Nontransmission of deoxynivalenol (vomitoxin) to milk following oral administration to dairy cows. *Journal of Environmental Science and Health*, 1984, B19, 593–609.
[108] Côté, L.M.; Dahlem A.M.; Yoshizawa I; Swanson S.P.; Buck W.B. Excretion of deoxynivalenol and its metabolite in milk, urine, and feces of lactating dairy cows. *Journal of Dairy Science*, 1986a, 69, 2416–2423.
[109] Prelusky, D. B.; Veira D.M.; Trenholm H.L. 1985. Plasma pharmacokinetics of the mycotoxin deoxynivalenol following oral and intravenous administration to sheep. *Journal of Environmental Science and Health,* 1985, B20, 603–624.
[110] Prelusky, D. B.; Veira D.M.; Trenholm H.L.; Hartin K.E. Excretion profiles of the mycotoxin deoxynivalenol, following oral and intravenous administration to sheep. *Fundamental and Applied Toxicology*, 1986, 6, 356–363.
[111] Prelusky, D. B.; Veira D.M.; Trenholm H.L.; Foster B.C. Metabolic fate and elimination in milk, urine, and bile of deoxynivalenol following administration to lactating sheep. *Journal of Environmental Science and Health*, 1987, B22, 125–148.
[112] Whitlow, L.M.; Nobel R.L.; Behlow R.F.; Hagler W.M.; Brownie F.G. Mycotoxins in North Carolina dairy feeds – a survey of 100 dairy farms. *Journal of Dairy Science,* 1986, 69,223-230.
[113] Speijers, G.J.; Speijers M.H. 2004. Combined toxic effects of mycotoxins. *Toxicology Letters*, 2004, 153, 91–98.
[114] Ingalls, J.R. Influence of deoxynivalenol on feed consumption by dairy cows. *Animal Feed Science and Technology,* 1996, 60, 297-300.
[115] Diaz, D.E.; Hagler W.M. Jr.; Hopkins B.A.; Patton R.A.; Brownie C.; Whitlow L.W. The effect of inclusion of a clay type sequestering agent on milk production of dairy cattle consuming mycotoxins contaminated feeds. *Journal of Dairy Science*, 2001, 84(abstr.), 1554.
[116] Acosta, Y.M. ; Mieres J.M. ; LaManna A. Effect of deoxynivalenol (DON) content of the concentrate on milk yield and milk quality. Abstract presented at the 21st Alltech Inc. annual symposium "Nutritional biotechnology in the feed and food industries". 2005. Lexington, KY.
[117] Windels, H.F.; DiCostanzo A.; Goodrich R.D. Effect of deoxynivalenol from barley on performance and health of large frame crossbred steers. *Minnesota Cattle Feeder Report,* B-417. St. Paul, MN, 1995.
[118] DiCostanzo, A; Meiske J.C.; Murphy M.; Epley L.; Felice L.; Johnson L.; Chester-Jones H. Effects of feeding vomitoxin-contaminated barley on performance of feedlot steers. *Minnesota Cattle Feeders Report.* St.Paul, MN,1995, B-418.
[119] DeHaan, K.; Stock R.; Brink D.; Klopkenstein T.; Schneider N. 1984. Scabby wheat influence on performance. *Nebraska Beef Cattle Report*, 1984, MP-47,6.
[120] Nelson, M., N.R. Schneider, A. R. Doster, M.P. Carlson, and T. Klopfenstein. 1984. Vomitoxin-contaminated wheat-pathology, toxicity in cattle. Nebraska Beef Cattle Report MP-47:3.
[121] Danicke S.; Gadeken D.; Ueberschar K.H.; Meyer U.; Scholz H. Effects of *Fusarium* toxin contaminated wheat and detoxifying agent on performance of growing bulls, on nutrient digestibility in wethers and on the carry over of zearalenone. *Arch Tierernahr,* 2002, 56, 245-61.

[122] Boland, E.W.; Anderson V.L.; Casper H.H.; Berg P.T.; Dhuyvetter D.V. The effects of increasing vomitoxin (DON) levels from scab infested barley fed to feedlot steers. Beef Prod. Field Day. Carrington Res. Ext. Ctr. Livestock Unit. North Dakota State University Ag. Expt. Sta, 1994, 17, 34.

[123] Harvey, R.B.; Kubena L.F.,Corrier D.E.; Witzel D.A.; Phillips T.D.; Heidelbaugh N.D. Effects of deoxynivalenol in a wheat ration fed to growing lambs. *American Journal of Veterinary Research*, 1986, 47, 1630-1632.

[124] Whitlow L.W.; Hagler W.M. Jr. Mycotoxin Toxicity, Treatment and Prevention. North *Carolina State University*. Raleigh, NC, 2004.

[125] Côté, L.M.; Beasley V.R.; Bratich P.M.; Swanson S.P.; Shivaprasad H.L.; Buck W.B. Sex-related reduced weight gains in growing swine fed diets containing deoxynivalenol. *Journal of Animal Science*, 1985, 61, 942-950.

[126] Keese, C; Meyer U.; Valenta H.; Schollenberger M.; Starke A.; Weber I.A.; Rehage J.; Breves G.; Danicke S. No carry over of unmetabolised deoxynivalenol in milk of dairy cows fed high concentrate proportions. *Molecular Nutrition and Food Research*, 2008, 52, 1514-29.

[127] U.S. Food and Drug Administration, Center for Food Safety and Applied Nutrition. 1993. Guidance for Industry and FDA: Letter to State Agricultural Directors; State Feed Control Officials, and Food, Feed and Grain Trade Organizations. Available at: http://www. cfsan.fda.gov/_dms/graingui.htmL. Accessed Apr. 26, 2006.

[128] Petrie, L.; Robb J.; Stewart A.F. The identification of T-2 toxin and its association with a hemorrhagic syndrome in cattle. *Veterinary Record*, 1977, 101, 326.

[129] Mirocha, C.J.; Pathre S.V.; Schauerhamer B.; Christiansen C.M. Natural occurrence of *Fusarium* toxins in feedstuffs. *Applied and Environmental Microbiology*, 1976, 32,553.

[130] Black, R.D.; McVet D.S.; Oehme F.W. 1992. Immunotoxicity in the bovine animal. *Veterinary and Human Toxicology,* 1992, 34, 438–442.

[131] Hsu, I.C,; Smalley E.B.; Strong F.M.; Ribelin W.E. Identification of T-2 toxin in moldy corn associated with a lethal toxicosis in dairy cattle. *Applied Microbiology,* 1972, 24, 682– 690.

[132] Kosuri, N.R.; Grave M.D.; Yates S.G.; Tallent W.H.; Ellis J.J.; Wolff I.A.; Nichols R.E. Response of cattle to mycotoxins of *Fusarium tricinctum* isolated from corn and fescue. *Journal of the American Veterinary Medical Association*, 1970, 157, 938.

[133] Beasley V.R.; Swanson S.P.; Corley R.A; Buck W.B.; Koritz G.D.; Burmeister H.R. Pharmacokinetics of the thrichothecene mycotoxin, T-2 toxin, in swine and cattle. *Toxicon*, 1986, 24, 13-23.

[134] Weaver, G.A.; Kurtz H.J.; Mirocha C.J.; Bates F.Y.; Behrens J.C.; Robison T.S.; Swanson S.P. The failure of T-2 toxin mycotoxin to produce hemorrhaging in dairy cattle. *The Canadian Veterinary Journal*, 1980, 21, 210.

[135] Cheeke, P.R. Mycotoxins in cereal grains and supplements. In: Cheeke, P.R. (Ed.), *Natural Toxicants in Feeds, Forages, and Poisonous Plants*. Interstate Publishers, Inc, Danville, IL, 1998a, 87–136.

[136] Mann, D.D.; Beuning G.M.; Hook B. 1983. Effects of T-2 toxin on bovine serum proteins. *American Journal of Veterinary Research*, 1983, 44, 1751–1759.

[137] Mann, D.D.; Buening G.M.; Osweiller G.D.; Hook B.S. Effect of subclinical levels of T-2 toxin on the bovine cellular immune system. *Canadian Journal of Comparative Medicine*, 1984, 43, 308–312.

[138] Buening, G.M.; Mann D.D.; Hook B.; Osweiller G.D. 1982. The effect of T-2 toxin on bovine immune system: cellular factors. *Veterinary Immunology and Immunopathology*, 1982, 3,411–418.

[139] Gentry, P.A.; Ross M.L.; Chan P. K-C. Effect of T-2 toxin on bovine hematological and serum enzyme parameters. *Veterinary and Human Toxicology*, 1984, 26, 24.

[140] Robinson, T.S.; Mirocha C.J.; Kurtz H.J.; Behren J.C.; Chi M.S.; Weaver G.A.; Nystrom S.D. Transmission o f T-2 toxin into bovine and porcine milk. *Journal of Dairy Science*, 1979, 62, 637-641.

[141] D'Mello, J.P.F.; MacDonald A.M.C. Mycotoxins. *Animal Feed Science and Technology*, 1997, 69, 155–166.

[142] Seeling K.; Danicke S.; Ueberscha K.H.; Lebzien R.P.; Flachowsky G. On the effects of *Fusarium* toxin-contaminated wheat and the feed intake level on the metabolism and carry over of zearalenone in dairy cows. *Food Additives and Contaminants*, 2005, 22, 847–855.

[143] Prelusky, D. B.; Scott P.M.; Trenholm H.L.; Lawrence G.A. Minimal transmission of zearalenone to milk of dairy cows. *Journal of Environmental Science and Health, Part B Pesticides, Food Contaminants and, Agricultural Wastes,* 1990, 25:87.

[144] CAST. *Mycotoxins, economic and health risks*. Task Force Report No.116. Council for Agricultural Science and Technology, Ames, IA, 1989.

[145] Weaver, G.A.; Kurtz H.J.; Behrens J.C.; Robison T.S.; Seguin B.E.; Bates F.Y.; Mirocha C.J. Effect of zearalenone on dairy cows. *American Journal of Veterinary Research*, 1986b, 47, 1826.

[146] Weaver, G.A.; Kurtz H.J.; Behrens J.C.; Robison T.S.; Seguin B.E.; Bates F.Y.; Mirocha C.J. Effect of zearalenone on the fertility of virgin dairy heifers. *American Journal of Veterinary Research*, 1986a, 47, 1395.

[147] Coppock, R.W.; Mostrom M.S.; Sparling C.G.; Jacobsen B.; Ross S.C. Apparent zearalenone intoxication in a dairy herd from feeding spoiled acid-treated corn. *Veterinary and Human Toxicology*, 1990, 32, 246-248.

[148] Towers, N.R.; Sprosen J.M.; Webber W. Zearalenone metabolites in cycling and non-cycling cows. In: *Toxinology and Food Safety*. Toxinology and Food Safety Research Group, Ruakura Research Centre, Hamilton, New Zealand, 1995a.

[149] Towers, N.R.; Wesselink C.; Fowke E.A.; Sprosen J.M. Plasma v*s* urinary zearalenone concentrations as indicators of zearalenone intake. In: *Toxinology and Food Safety*. Toxinology and Food Safety Research Group, Ruakura Research Centre, Hamilton, New Zealand, 1995b.

[150] Smith, J.; Wesselink C.; Parr J.; Sprosen J.M.; Fowke E.A.; Towers N.R.; Laboyrie D. Effect of Zearalenone on ewe pregnancy rates. In: *Toxinology and Food Safety*. Toxinology and Food Safety Research Group, Ruakura Research Centre, Hamilton, New Zealand, 1995.

[151] Sprosen, J. M.; Towers N.R. Urinary Zearalenone metabolite concentrations in herds with fertility problems. In: *A Toxinology and Food Safety*. Toxinology and Food Safety Research Group, Ruakura Research Centre, Hamilton, New Zealand, 1995.

[152] Kellela, K.; Ettala E. The oestrogenic *Fusarium* toxin Zearalenone in hay as a cause of early abortions in the cow. *Nord Veterinary Medicine*, 1984, 36, 305.

[153] Mirocha, C.J.; Schauerhamer B.; Pathre S.V. Isolation, detection and quantitation of Zearalenone in maize and barley. *Journal of Association of Official Analytical Chemists*, 1974, 57, 1104.

[154] Dicostanzo, A.; Johnston L.W.H.; Murphy M. A review of the effects of molds and mycotoxins in ruminants. *Journal of Animal Science*, 1996, 12, 138–150.

[155] Murata H.; Shimada N.; Yoshioka M. Supressive effect of zearalenone, an estrogenic mycotoxin, on bovine neutrophil chemiluminescence. *Veterinary and Human Toxicolog.*, 2002, 44, 83-86.

[156] Prelusky, D.B.; Savard M.E.; Trenholm H.L. Pilot study on the plasma pharmacokinetics of fumonisin B_1 in cows following a single dose by oral gavage or intravenous administration. *Journal of Natural Toxins,*1995, 3, 389-394.

[157] Prelusky, D.B.; Trenholm H.L.; Rotter B.A.; Miller J.D.; Savard M.E.; Yeung J.M.; Scott P.M. Biological fate of fumonisin B_1 in food-producing animals. *Advances in Experimental Medicine and Biology*, 1996, 392,265-278.

[158] Richard J.L.; Meerdink G.; Maragos C.M.; Tumbleson M.; Bordson G.; Rice L.G.; Ross P.F. Absence of detectable fumonisins in the milk of cows fed *Fusarium proliferatum* (Matsushima) Nirenberg culture material. *Mycopathologia.*, 1996, 133, 123–126.

[159] Scott, P.M.; Delgado T.; Prelusky D.B; Trenholm H.L.; Miller J.D. 1994. Determination of fumonisin in milk. *Journal of Environmental Science and Health*, 1994, B29, 989-998.

[160] Smith J.S.; Thakur R.A. Occurance and fate of fumonisins in beef. *Advances in Experimental Medicine and Biology*, 1996, 392, 39-55.

[161] Diaz, G.J.; Boermans H.J. Fumonisin toxicosis in domestic animals: a review. *Veterinary and Human Toxicology*, 1994, 36, 548–555.

[162] Baker D.C.; Rottinghaus G.E. Chronic experimental fumonisin intoxication of calves. *Journal of Veterinary Diagnostic Invesigation*, 1999, 11, 289–292.

[163] Mathur, S.; Constable P.D.; Eppley R.M.; Waggoner A.L.; Tumbleson M.E.; Haschek W.M. Fumonisin B_1 is hepatotoxic and nephrotoxic in milk-fed calves. *Toxicology Science,* 2001, 60, 385–396.

[164] Diaz, D.E.; Hopkins B.A.; Leonard L.M.; Hagler W.M., Jr.; Whitlow L.W. Effect of fumonisin on lactating dairy cattle. *Journal of Dairy Science*, 2000, 83(abstr.),1171.

[165] Osweiler, G.D.; Kehrli M.E.; Stabel J.R.; Thurston J.R.; Ross P.F.; Wilson T.M. Effects of fumonisin-contaminated corn screenings on growth and health of feeder calves. *Journal of Animal Science*, 1993, 71, 459-466.

[166] Edrington T.S.; Kamps-Holtzapple C.A.; Harvey R.B.; Kubena L.F.; Elissalde M.H.; Rottinghaus G.E. Acute hepatic and renal toxicity in lambs dosed with fumonisin-containing culture material. *Journal of Animal Science*, 1995, 73, 508-515.

[167] Federal Register. Guidance for industry: Fumonisin levels in human foods and animal feeds. *Federal Register 66.*, 2001, 218, 56688-56689.

[168] Marquardt R.R.; Frohlich A.A. A review of recent advances in understanding ochratoxicosis. *Journal of Animal Science*, 1992, 70, 3968-3988.

[169] Blank R.; Rolfs J.P.; Sudekum K.H.; Frohlich A.A.; Marquart R.R.; Wolffram S. Effects of chronic ingestion of ochratoxin A on blood levels and excretion of the mycotoxin in sheep. *Journal of Agrucultural and Food Chemistry,* 2003, 51, 6899-6905.

[170] EFSA. Opinion of the Scientific Panel on Contaminants in Food Chain on a request from the Commission related to ochratoxin A (OTA) as undesirable substances in animal feed. *European Food Safety Authority*, 2004b, EFSA-Q-2003-039.

[171] Griffiths, I.B.; Done S.H. Citrinin as a possible cause of thepruritis, pyrexia, haemorrhagic syndrome in cattle. *Veterinary Research*, 1991, 129, 113–117.

[172] Patterson, D.S.P.; Shreeve B.J.; Roberts B.A.; Berrett S.; Brush P.J.; Glancy E.M. Effect on calves of barley naturally contaminated with ochratoxin A and groundnut meal contaminated with low concentration of aflatoxin B_1. *Research in Veterinary Science*, 1981, 31, 213–218.

[173] Lioi M.B.; Santoro A.; Barbieri R.; Salzano S. ; Ursini M.V. Ochratoxin A and zearalenone: a comparative study on genotoxic effects and cell death induced in bovine lymphocytes. *Mutation Research*, 2004, 557, 19-27.

[174] Carlton W.W.; Krogh P. Ochratoxins. In: Shimoda W. (Ed.). In Conference on *Mycotoxins in animal feeds and grains related to animal health*, Report FDA, Rockville, MD, USA, 1979, 165-287.

[175] Brietholtz-Emanuelsson A.; Olsen M.; Oskarson A; Palminger I; Hult K. Ochratoxin A in cow's milk and in human milk with corresponding human blood samples. *Journal AOAC International,* 1993, 76, 842-846.

[176] Skaug, M.A. Analysis of Norwegian milk and infant formulas for ochratoxin A. *Food Additives and Contaminants*, 1999, 16, 75–78.

[177] FCA. Survey of milk for mycotoxins. *Food Standards Agency*, 2001, 17/01

[178] Burke, J.M.; Rorie R.W.; Piper E.L.; Jackson W.G. Reproductive responses to grazing endophyte-infected tall fescue by post-partum beef cows. *Theriogenology,* 2001, 56, 357–369.

[179] EFSA. Opinion of the Scientific Panel on Contaminants in Food Chain on a request from the Commission related to ergot as undesirable substance in animal feed. *European Food Safety Authority,* 2005, 225, 1–27.

[180] McMullen M.; Stoltenow C. Ergot. North Dakota State University, 2002, 551.

[181] Aellig, W.H.; Nuesch E. Comparative pharmacokinetic investigations with tritiumLabeled ergot alkaloids after oral and intravenous administration in man. Int. *Journal of Clinical Pharmacology* 1977, 15, 106-112.

[182] Whittemore, C.T.; Macer R.C.F.; Miller J.K.; Mantle P.G. Some consequences of the ingestion by young and growing pigs of feed contaminated with ergot. *Research in Veterinary Science*, 1976, 20, 61-69.

[183] Whittemore, C.T.; Miller J.K.; Mantle P.G. Further studies concerning the toxicity of ingested ergot sclerotia (*Claviceps purpurea*) to young and growing pigs. *Research in Veterinary Science*, 1977, 22, 146-150.

[184] Lorenz, K. Ergot on cereal grains. *CRC Critical Reviews in Food Science and Nutrition*, 1979, 11, 311-354.

[185] Tfelt-Hansen, P.; Saxena P.R.; Ferrari M.D. Ergot alkaloids. *Handbook of Clinical Neurology,* 1995, 21, 61-78.

[186] Filipov, N.M.; Thompson F.N.; Tsunoda M.; Sharma R.P. Region-specific decrease of dopamine and its metabolites in brains of mice given ergotamine. *Journal of Toxicology and Environmental Health*, 1999, 56, 47-58.

[187] Mantle, P.G. Ergotism in cattle, sheep, swine. In: Wyllie, T.D. and L.C. Morehouse, ed. *Mycotoxic fungi, mycotoxins, mycotoxicoses: an encyclopedia handbook*. New York, Marcel Dekker, 1977, 145-151, 207-213, 273-275.

[188] Howard, M.D.; Muntifering R.B.; Bradley N.W.; Mitchell G.E.; Lowry S.R.Jr. Voluntary intake and ingestive behavior of steers grazing Johnstone or endophyte-infected Kentucky-31 tall fescue. *Journal of Animal Science*, 1992, 70, 1227–1237.

[189] Cheeke, P.R. Mycotoxins associated with forages. In: Cheeke, P.R. (Ed.), *Natural Toxicants in Feeds, Forages, and Poisonous Plants*. Interstate Publishers, Inc, Danville, IL, 1998b, 243–274.

[190] Woods A.J; Bradley-Jones J.; Mantle P.G. An outbreak of gangrenous ergotism in cattle. *Veterinary Records,* 1966, 78,742-749.

[191] Blaney, B.J.; McKenzie R.A.; Josey B.J.; Ryley M.J.; Downing J.A. Effect of grazing sorghum (sorghum bicolour) infected with ergot (*Claviceps africana*) on beef cattle. *Australian Veterinary Journal*, 2000, 78, 124-125.

[192] Appleyard, W.T.Outbreak of bovine abortion attributed to ergot poisoning. *Veterinary Research*, 1986, 118, 48-49.

[193] Ross, A.D.; Bryden W.L.; Bakua W.; Burgess L.W. Induction of heat stress in beef cattle by feeding the ergots of *Claviceps pupurea*. *Australian Veterinary Journal*, 1989, 66, 247–249.

[194] Loken T. Ergot from meadow grass in Norway-chemical composition and toxicological effects in sheep. *Nord.Veterinary Medicine*, 1984, 36, 259-65.

[195] Parkheava. Zit. nach: Wolff J,; Richter W.I.F.; Spann B. 1995. *Mutterkornalkaloide in der Milch?* VDLUFA-Schriftenreihe, Kongressband. 1979, 18, 515-524.

[196] Wolff, J.; Richter W.I.F.; Spann B. Mutterkornalkaloide in der Milch? *VDLUFA Schriftenreihe*, Kongressband, 1995, 18, 515-524.

In: Mycotoxicoses in Animals Economically Important
Editors: Edlyane Gonçalez, et. al. pp. 67-95
ISBN: 978-1-61668-195-1

Chapter 3

MYCOTOXICOSIS IN SWINE

***Carlos A. Mallmann*[1]*, Paulo Dilkin*[1*]*, Ricardo Hummes Rauber*[2] *and Leandro Zanini Giacomini*[2]**

[1*]Departament of Preventive Veterinary Medicine
Universidade Federal de Santa Maria Prédio 44 - Asa Norte, 4. andar
CEP 97.105-910, Santa Maria (RS) BRAZIL
[2] Instituto SAMITEC
Rodovia RSC 287, km 227, nº. 2800 – Faxinal da Palma
CEP 97.105-910, Santa Maria (RS) BRAZIL

ABSTRACT

Mycotoxins are secondary metabolites produced by several filamentous fungi ubiquitarily found in the nature on the alimentary substrate that composes the diet of swine. In tropical and subtropical climates, the fungal development is favored by factors as high humidity condition and temperature. More than five hundred mycotoxins produced by approximately a hundred of fungi are known so far. The main ones can be divided into three groups: aflatoxins produced by fungi of the genus *Aspergillus*, as *A. flavus* and *A. parasiticus*; ochratoxins produced by fungi of the genus *Aspergillus* and *Penicillium*; and fusariotoxins whose main representatives are the trichothecenes, zearalenone and fumonisins, produced by several species of the genus *Fusarium*. When ingested through human or animal feeding, mycotoxins can produce several harmful effects to the consumer's health, implicating in enormous economical, sanitary and commercial damages, above all for their anabolizant, estrogenic, carcinogenic, mutagenic and theratogenic properties, besides other less poisonous effects, also harmful to the health and the productivity. The main mycotoxins and target organs in swine are: aflatoxins in the liver; zearalenone in the reproductive system; ochratoxin A in kidneys; fumonisins in lungs and trichothecenes in the digestive tract. The emergence of toxicosis signs is intimately related to the dose and time of consumption of each toxin. Approximately 90% of the toxicosis are chronic and do not show specific clinical signs.

* Corresponding author: P. Dilkin
Telephone/Fax: + 55 55 3220 8445
E-mail address: dilkin@lamic.ufsm.br

They could be easily confused with malnutrition, deficient manage or other chronic diseases that implicate in the decrease of the productivity of animals. Therapeutic methods have low efficiency in reducing the impact of the toxicosis. The presumptive diagnosis of mycotoxicosis is based on the observation of the clinical signs of the intoxicated animals and the analysis of environmental data regarding the crop and to the storage of cereals used in feeding of swine. The diagnosis is accomplished by the analysis of the presence of the mycotoxin in the feed of the intoxicated animals. The more used techniques are analyses through ELISA kits, Thin-Layer Chromatography (TLC) and High-Performance Liquid Chromatography (HPLC). The use of diagnostic methods like Mass Spectrometry (GC/MS, LC/MS and LC/MS/MS) is recent. However, it is already known that those are the more efficient methodologies and that will substitute the others in a very close future. Preventive measures, as practices that inhibit fungi growth or the use of anti-mycotoxins addictive in the feeding of the animals present good effectiveness.

INTRODUCTION

Mycotoxicosis is the toxicosis of animals by mycotoxins, substances resulting from the secondary metabolism of filamentous fungi [1]. Clinical forms of mycotoxicosis can be acute, less frequent, or chronical, more than 90% of it [2, 3]. There are no specific clinical signs and it could be easily confused with malnutrition, deficient manage practices or other chronic diseases that implicate in the decrease of productivity of the animals. Acute mycotoxicosis is characterized by the manifestation of specific clinical signs, inducing acute disease, and culminating with death of animals [3].

Mycotoxins are organic compounds of low molecular weight and low immunogenic capability. They occur worldwide, however they prevail in tropical and subtropical climates, where fungal development is favored by environmental conditions. Main species of fungi producing mycotoxins belong to the genus *Aspergillus*, *Penicillium*, *Fusarium*, *Claviceps*, *Alternaria*, *Pithomyces*, *Mirothecium*, *Stachybotrys* and *Phoma* [1]. In tropical and subtropical climates, the fungal development is favored by factors as excellent humidity conditions and temperature [1]. More than five hundred mycotoxins produced by approximately one hundred fungi are known nowadays. The main ones can be divided into three groups: aflatoxins, produced by fungi of the genus *Aspergillus* as *A. flavus* and *A. parasiticus*; ochratoxins, produced by fungi of the genus *Aspergillus* and *Penicillium* and fusariotoxins that possess as main representatives the trichothecenes, zearalenone and fumonisins, produced by several species of the genus *Fusarium* [3].

Whenever the ingestion of mycotoxins occurs by humans or animals several harmful effects can be produced to the consumer's health, involving huge damages of economical, sanitary and commercial orders, especially for their anabolizant, estrogenic, carcinogenic, mutagenic and theratogenical properties beside other less poisonous effects, and also harmful to the health and productivity [1, 4]. Nowadays, the knowledge on the physiology, production of toxins and development of the fungi producing mycotoxins is relatively restricted. However, it is known that they grow and proliferate well in cereals and grains, especially in the peanut, corn, wheat, barley, sorghum and rice, in which usually there are highly nutritious substrate for their development [3]. Cereals and grains, in turn, considerably lose nutritional qualities, besides being contaminated with mycotoxins that can remain for several years in substrates, even in the absence of producer fungi [1]. Fungal growth and production of mycotoxins in cereals may occur in various stages of development, maturation, harvesting, transport, processing or storage of grain [3, 5]. The difficulty of harvesting cereal in the correct stage of humidity and maturation may be one of the most critical point regarding the formation of mycotoxins. It is natural that cereals harvested with high humidity facilitate fungal development before drying, or even in the warehouse, especially when they are not submitted to appropriate drying process. Furthermore, cereals in ploughing, after the physiological maturation of the grains, shall be subject to high humidity and pests attack that provide favourable condition to the fungal development. The transport of cereals with high humidity by long distances or prolonged time causes the development of fungi by the formation of an appropriate microclimate. Therefore, the reduction of the cereals' moisture through drying inside dryers is of fundamental importance [3, 5].

Drying, in turn, when poorly executed may, in many ways, contribute to the proliferation of fungi in grains. The long queues and the long wait time for cereal haulier trucks on port storage units often motivate acceleration of drying process, by the increase on driers temperature, causing injury to the grain and the consequent decrease in its protection. In addition, a series of shortcomings in the store, such as: cereals stored with high humidity; impurity; poor distribution of cereals in silo; ventilation failure and incidence of pests, are factors that often led to the fungal development and production of mycotoxins. Understanding of a number of factors which may lead to the formation of these toxic metabolites is vital for intervention in the process, in order to diminish such production of mycotoxins [3].

The various toxic effects are due to different chemical structures of mycotoxins, influenced by being swallowed by different animal organisms and by the diversity of species, breed, sex, age, environmental factors, nutritional conditions and the presence of other chemicals [6].

Despite a series of variations, an estimate of susceptibility to mycotoxicoses can be made. For that, it is necessary to quantify the level of contamination of feed, associated with the type and severity of diseases related to its consumption. In this way, mycotoxicoses can be divided into acute and chronic. The acute manifestations occur when individuals consume moderate to high doses of mycotoxins. Clinical signs may appear together with specific pathological signs, depending on the mycotoxin ingested, the susceptibility of the specie, individual body condition and interaction with other factors, including other mycotoxins. Lesions depend on each mycotoxin, but the main ones relate to liver damage, bleeding, nephritis, digestive mucosal necrosis and death [3, 7- 9].

The chronic mycotoxicosis occurs when there is a moderate to low doses are consumed. In such cases, acute manifestations of toxicosis generally do not occur. However, the picture is characterized by the reduction of reproductive efficiency, feed consumption, rate of growth and weight gain. This picture is only detected with special care or through a programme for monitoring of mycotoxins on feed. Clinical signs can still be confused with other diseases arising from mycotoxicosis or with nutritional deficiencies [10].

There are few available statistics regarding the incidence of mycotoxicosis, but there is a general awareness that *the hidded danger* (chronic toxicosis) is responsible for most of losses in flocks. There are still effects caused especially by chronic mycotoxicosis that can lead to immunosuppression, leaving the individual predisposed to other diseases whose pathogens easily multiply with the fall of host resistance. On the other hand, the lack of an adequate response to vaccination is often resulting from the ingestion of mycotoxins [3].

The presumptive diagnosis of mycotoxicosis is based on observation of clinical signs of intoxicated animals and environmental data analysis regarding the harvest and storage of grain used in swine feed. Typically, the story of the introduction of a new feed, sometimes with physical characteristics altered, is associated with the context of toxicosis. However, the definitive diagnosis is performed through the analysis of the presence of mycotoxin in feed of intoxicated animal [10]. The most used techniques are analyses by ELISA kits, thin-layer chromatography (TLC) and high-performance liquid chromatography (HPLC). The use of diagnostic methods of gas and liquid chromatography-mass spectrometry (GC/MS, LC/MS and LC/MS/MS) is recent. However, it is already known that these are the most efficient and secure methodologies and will replace the others in a very near future [3, 11].

The treatment of mycotoxicosis constitutes one of the biggest challenges in the veterinary clinic. The withdrawal of contaminated feed is the first measure to be adopted and the supporting treatment seems to improve a little the prognosis. Adding greater levels of sulphur amino acids in feedstuffs is adopted by veterinarian practitioners; however its effectiveness needs greater scientific studies [3].

The main prophylactic measures consist on adopting culture techniques and management that make fungal growth difficult, such as cereals harvest immediately after the physiological maturation, leaving cereals less exposed to weather conditions, and drying and stocking in warehouses suitable for each type of cereal or by-product. The use of organic acids can help in conservation of feed when in situations of risk and the use of additives, either natural or modified by adding enzyme or biological compounds for feed, deserves further scientific, but in some situations and experimental field, some promising results have been shown for swine. Constant and continuous monitoring of mycotoxins in the production of feed, including sampling techniques appropriate to the detection of the problem, is the most efficient technical option and, to the economical point of view, more viable, especially for the swine industry of medium and large sizes. Anti-mycotoxins additives (adsorbents) can and, sometimes, must be used in contaminated feed [3, 12].

AFLATOXINS

Aflatoxins are secondary metabolites of strains of toxigenic fungi, of the genus *Aspergillus*, especially *A. flavus and A. parasiticus*, belonging to the group of the bifuranocoumarins. They are formed by heterocyclic molecules with atoms of oxygen and ring of difurano, which differ ones to others only by small variations in its basic molecular structure [1]. Among the hundreds of well-known mycotoxins, aflatoxins are considered the most important in Brazil [2, 3]. Many compounds have been described as being aflatoxins; however, only aflatoxins B_1, B_2, G_1 and G_2 were identified as natural contaminants of seeds and feed. Other aflatoxins have been isolated from milk, meat and urine of animals, as it is the case of aflatoxins M_1 and M_2, which are metabolites of aflatoxins B_1 and B_2, respectively. These mycotoxins are produced in the liver of animals few hours after ingestion of contaminated feed. Typically, aflatoxin B_1 presents the greatest prevalence in various agricultural products, and being also the most toxic of the group [3].

Occurrence of Aflatoxins in Feed

Evaluations carried out by the Laboratory of Mycotoxins at the *Universidade Federal de* Santa Maria (LAMIC/UFSM) in Brazil, during the recent 16 years have shown that aflatoxins in maize was present in 49.06% of the 51,553 samples tested in the routine of the laboratory, with an average of contamination 10.3 μg/kg. The general positivity was 38.03% of the 108,678 analysed samples, with an average of contamination of 9.03 μg/kg, slightly less than that of corn with the inclusion of raw materials and feed with lower average concentration of aflatoxins [13].

Clinical Signs and Lesions

The clinical signs and the seriousness of aflatoxicosis may vary due to the age of animals (younger or foetuses have more sensitivity), type of aflatoxin (B_1 has greater toxicity), composition of diet (diets with high levels of protein protect partially against toxic effects), toxin exposure time, nutritional status and, above all, due to aflatoxins' concentration on feed. Thus, the aflatoxicosis occurs as two clinical forms: acute or chronic. Both are dose-dependent and dependent on exposure time. The acute toxicosis is readily recognized thought liver lesions, and clinically characterized by depression, anorexia, jaundice and bleeding. In the acute cases, clinical signs are evidenced in approximately 6 hours after ingestion and animal quickly lead to death. In less serious cases, where clinical signs begin between 6 and 12 hours, increased body temperature (40.0 the 41.1°C) can be detected accompanied by muscular tremors and innapetence, incoordenation caused by haemorragies that occur in muscles of the shank. Diarrhoea can be present and poorly digested feed, dark green in colour and generally liquid (Figure 1). In lactating females it is common to observe fast involution of the mammary gland and the poor state of piglets. After 24 hours, pigs may show haemorrhagic diarrhoea and diphtheria. The pigs that outperform acute aflatoxicosis picture, usually not retrieve more weight lost nor the aspect of healthy animal. They remain slim, with bad feed convertion rate even after great period after the toxicosis [3, 7-9].

The main economic losses result from the ingestion of low concentrations of aflatoxins in feed, which leads to chronic toxicosis, representing more than 90% of cases. Thus, the biggest concerns among the producers are due to sub-clinical toxicosis, because the level of aflatoxins found in feed is often insufficient to trigger a noticeable clinical picture, but induces significant losses to the flock productivity.

The chronic aflatoxicosis occurs when toxins are ingested in small doses and concentrations, less than 1 mg/kg of feed, for extended periods of time. In such cases, there are only inappetence, lethargy, reduced weight gain and bad general appearance and sometimes diarrhea occurs few weeks after the beginning of the consumption of contaminated feed. However, the clinical signs present a slow evolution. Scaly skin, erected hair and reddish-purple coloured areas in the skin, lethargy and depression are noted. This form also is perceived by the decrease in feed consumption, low weight gain, deteriorating feed conversion rate and bad productive performance of animals. With progression to the final stages, signs of ataxia and jaundice frequently occur and sometimes convulsions. Because of the delay in the perception of these consequences and natural difficulties for clinical diagnosis, economic losses may be particularly severe. Furthermore, chronic toxicosis often is not properly diagnosed, being confused with management problems, nutritional deficiencies, and animal genetic quality or even assigned to the incidence of opportunistic infections resulting from chronic aflatoxicosis itself [3, 4, 14].

Figure 1. Diarrhea and poorly digested feed from pig intoxicated with aflatoxins.

Figure 2. Differences on size of animals intoxicated with aflatoxins.

Due to differences on the sensitivity that exists among the animals, even when the lot is formed by animals of the same descendence, differences in the development is almost always present, forming the so-called "unpairement on the lot" (Figure 2) as weight and size of animals. This clinical form induces immunosuppression; it is highly mutagenic, theratogenic and carcinogenic, moreover inducing hepatotoxicosis, with bile ducts proliferation, hepatic steatosis and haemorrhages. In more severe cases in which the disease reaches the terminal stages, pathological changes are characterized by jaundice in the mucous membranes and subcutaneous tissue as well as diffuse haemorrhages, especially in the kidneys and digestive system. Intra-abdominal and/or thoracic haemorrhages may occur together with ascites [2-4].

Although the literature present several citations saying that chronic toxicosis in pigs occur at concentrations of toxin below 1 mg/kg, there is a consensus in the scientific medium, allied to findings during recent years, that intake levels as low as 10 µg of total aflatoxins per kg of feed provided to pigs can already cause significant productivity losses. These effects are most frequently observed in the decrease of feed consumption and consequently of the gain weight and decrease in average number of piglets born by sow/year [3].

Clinical Pathology

Liver lesions, more than producing a decreased protein synthesis, show increased activity of the enzymes aspartate aminotransferase, gamma glutamiltransferase and alkaline phosphatase. Increased liver enzyme activity as arginase, sorbitol dehydrogenase and alanina aminotransferase is an indication of liver injury in various species, having, therefore, an important diagnostic value. Sorbitol dehydrogenase is an enzyme of cell cytoplasm indicative of liver injury. Its persistency in the circulation occurs for a short period of time, and thus replaced in the diagnostic examination by alanine aminotransferase. The time of prothrombin, which is approximately 20 seconds in swine, increases to approximately 60 seconds between 12 to 24 hours after the toxicosis. Still, as a consequence of liver lesion, a decrease in the amount of plasma proteins is always present after a period of toxicosis [3, 10].

Macroscopic Lesions

In acute aflatoxicosis, animal death usually occurs over a period of 3 days [10]. Changes immediately evident to external examination are the pallor of mucous membranes and bulging of the abdomen, because of the reduction in feed intake, dehydration and gastric emptying caused by diarrhoea. The presence of liquid faeces can usually be found in the perineal zone because undigested feeds are eliminated in abundance, even during the period in which the animal is already dying in decubit pad [7, 15].

The liver is always the most affected organ, and its changes are related to the time elapsed since the beginning of toxicosis. After 6 hours of acute toxicosis, liver presents a pale-brownish colour with cooked appearance. According to the evolution of the disease, the liver presents haemorrhagic foci with approximately 1 mm in diameter on the surface, and then shows dark pale colour, edematous and friable (Figure 3). The gallbladder presents thickening of the walls, induced by oedema with fluid in the submucosa, muscular and serous layers [10].

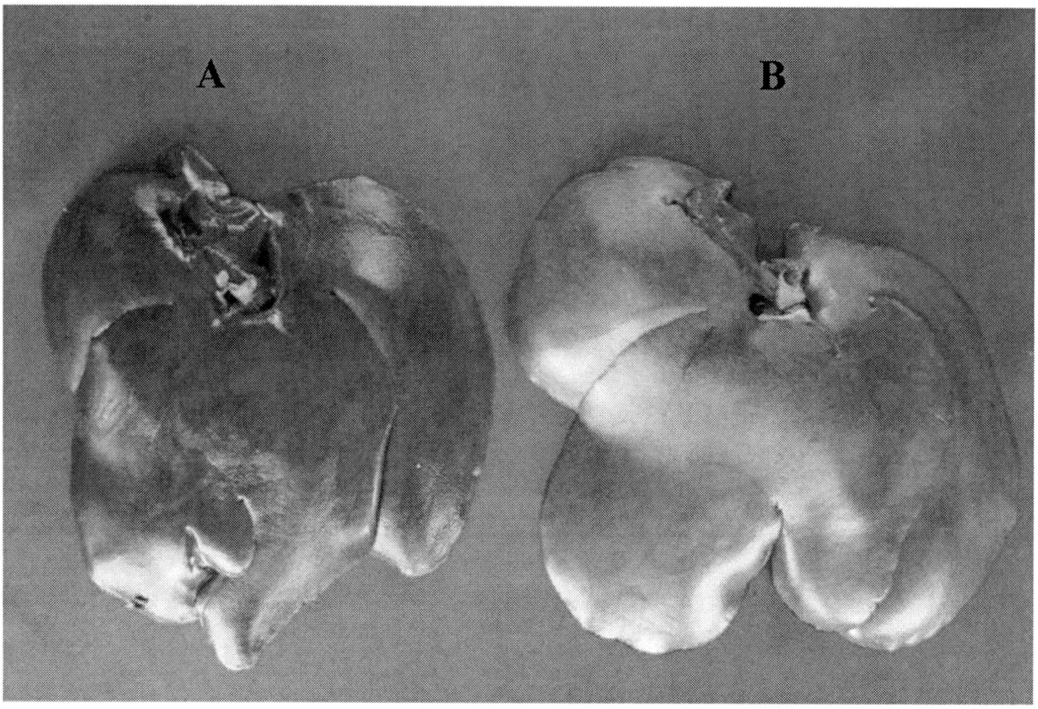

Figure 3. (A) normal liver; (B) liver from pig intoxicated with aflatoxins.

The heart can have haemorrhagic areas, affecting mainly the subepicardic and endocardic regions. An extensive enterocolitis may occur across intestinal lumen, but with greater frequency in the jejunum and ileum, where disseminated haemorrhages with free blood in the lumen can be seen. Arterial hyperaemia in the cecum and colon are common and may also contain varying amounts of free blood in the lumen [10].

Acute toxicosis can still induce formation of liquid collections around organs in different cavities. Most often, these collections are observed in heart and peritoneum. The jaundice in various organs and tissues can be observed, especially when toxicosis passes to subacute or to chronic form. The icteric carcase (associated to edematous and yellowish liver) is an important indicator of toxicosis.

Chronic exposures to low doses of aflatoxin result in jaundice throughout the body. The liver may show yellowish pale aspect with well-defined lobed with haemorrhagic foci, especially across parietal surface. It always displays different degrees of fibrosis and/or cirrhosis. In case of cirrhosis, a clear liquid (straw yellow) can be found in the abdominal and thoracic cavities, with lung oedema, induced by liquid buildup on the organ [3, 9]. Under these conditions, the gallbladder is usually oedematous, increased in volume, and the content thicker than it is typically found in healthy animals. In such cases, jaundice tends to be generalized, accompanied by yellowish liquids and fluids, oedemas from varied location, targeting in particular the thoracic and abdominal cavities, especially in the spirals of the colon. In the thoracic cavity, there is a predilection to accumulate yellowish liquid in pericardial sac [10]. The decrease in the blood coagulation time also occurs, and therefore haemorrhagic liquid collections in cavities and in mucous membranes and haemorrhages in muscular mass can be seen [3, 15}.

Kidneys are usually oedematous or hypertrofic with pale colour, according to the seriousness of toxicosis. Some researchers also report the possibility of areas of consolidation in the lung parenchyma [3, 10].

The LD_{50} of aflatoxins for pigs is very low (0.62 mg/kg), and the considered safe dose is 50 μg/kg (ppb). However, in situations of field, due to the difficulty to correctly sample batches of feed, this level of toxin is considered to be very high. Lower levels remain important in the induction of immunosuppression, which sometimes leads to the difficulty on establishing a clinical correlation with aflatoxin toxicosis [10].

Effects of Aflatoxins on the Immunologic System

There is a consensus that the mechanism of the immunotoxicity is not clear. The mycotoxins effects on the complement system, interferon and seric protein concentration are primarily resulting from the liver damage and diminished production of proteins. More than harming the formation of interferon and complement, it is known that aflatoxins diminish the phagocytic capacity of macrophages and migration of lymphocytes and leukocytes. In addition, they cause aplasia of thymus and significant decline of weight of Fabricius bursa in poultry, affecting more the cellular immunity. Aflatoxin B_1 affects T lymphocytes, including both T-helper as well as T-suppressor cells [2, 3, 10].

The effects of aflatoxins over the immunoglobulins also are not clear. Various studies indicate a decrease in levels of IgA and IgG antibodies in animals which are intoxicated with feed containing more than 500 mg/kg of aflatoxins. The same importance of aflatoxins due to the clinical effects in pigs can also be observed by the decrease of immunity and secondary diseases increase in pigs severely affected, especially caused by *Salmonella*, *Pasteurella* and *Erysipelothrix*. Failures or decrease in vaccine response in different species have been often related to the incidence of aflatoxins in diet and clinical manifestations of aflatoxicosis in the flock [3].

Diagnosis of Aflatoxins

Clinical signs of acute aflatoxicosis do not represent a greater difficulty for the diagnosis. Commonly, the story of the introduction of a new batch of feed, often with gross altered characteristics is associated with the picture. The clinical signs and pathologic-anatomical findings suggest aflatoxin toxicosis. The greater difficulty is due to the time for the trigger clinical signs, because, due to the usually found concentrations in commercial rations, it may take two weeks or more. The analysis of feed through appropriate sampling system is the only safe way to perform the clinical diagnosis [3, 6].

ZEARALENONE

Zearalenone is a fungal estrogenic metabolite non-steroid, chemically described as a lactone of the resorcilic phenic acid. It has good thermal stability and low solubility in water; however, it has high solubility in organic solvents. It is produced by several species of fungi of the genus *Fusarium* including *F. culmorum*, *F. graminearum* and *F. crookwellense* [1]. These species colonize cereals and tend to become particularly important during seasons of high humidity, accompanied by mild temperatures. Fluctuations with low temperatures are suitable for the production of large quantity of mycotoxin. Temperatures in the range of 25°C is favorable to the fungal growth, while its reduction for approximately 10°C in the presence of higher humidity than 17%, triggers the secondary metabolism, responsible for the production of zearalenone. For this reason, it is a natural contaminant mycotoxin of various cereals, such as wheat, barley, rice and, particularly in corn, in several countries of the world [1, 3, 13].

About the occurrence of zearalenone in Brazil, the results during the recent 16 years have shown that zearalenone of 88,579 random samples taken by the feed industry and analysed in LAMIC/UFSM showed contamination of 20.09% of them. Although the average of positivity is low, at certain times of the year the limit rate is exceeded in 45% of the samples analysed. In this laboratory, the average concentration of zearalenone found was 49.8 μg/kg, and the maximum level detected was as high as 18.5 mg/kg [13].

Clinical Signs

Swine seem to be the most sensitive species to zearalenone. Zearalenone and/or products of its biotransformation are perpetrators of hyperestrogenism syndrome in pigs, medically known as vulvovaginitis picture (Figure 4). Clinical signs presented by this toxicosis vary according to the dose of toxin ingested and the age of animals, but they are essentially characteristic. Thus, the pictures tend to be more acute as younger would be the animals and as higher would be the dose of ingested toxin. Therefore, the young gilts, with approximately 4 months of age, are usually the most affected. However, the beginning of the presentation of the first clinical signs after ingestion of contaminated feed varies greatly, taking usually between one to four weeks [3, 16].

High sensitivity of young animals has already been established in various natural poisoning, in which they showed characteristic clinical signs, without, however, affecting adult animals fed with the same feed containing the same concentrations of zearalenone. Sick animals show signs that mimetize the estrous, like the reddening and increased vulvar lips and swelling of mammary glands with significant increase in volume (figure 4). Sometimes, the animals may show anxiety, resulting in fights or cannibalism. Clinical manifestations are usually persisted for long periods, even after the replacement of contaminated feed. Often the females do not return to cycling again. The vaginal or rectal prolapse occurs because of the sphincters' relax, but it is a signal not always present in an intoxicated flock, particularly in males, and sometimes they can be the only ones with obvious clinical signs of poisoning by zearalenone. In such cases, usually secondary infection occurs as a result of injuries and contamination that occur on the surface of the mucosa of these organs [2].

Oedema of the mammary gland in adult sows often leads to suspect of pregnancy or proximity of parturition. The increase in volume of the vulva and mammary gland is more evident in the last two weeks of pregnancy of poisoned animals, and the presence of milk in the mammary gland can be anticipated in at least two weeks [2, 3, 16-18].

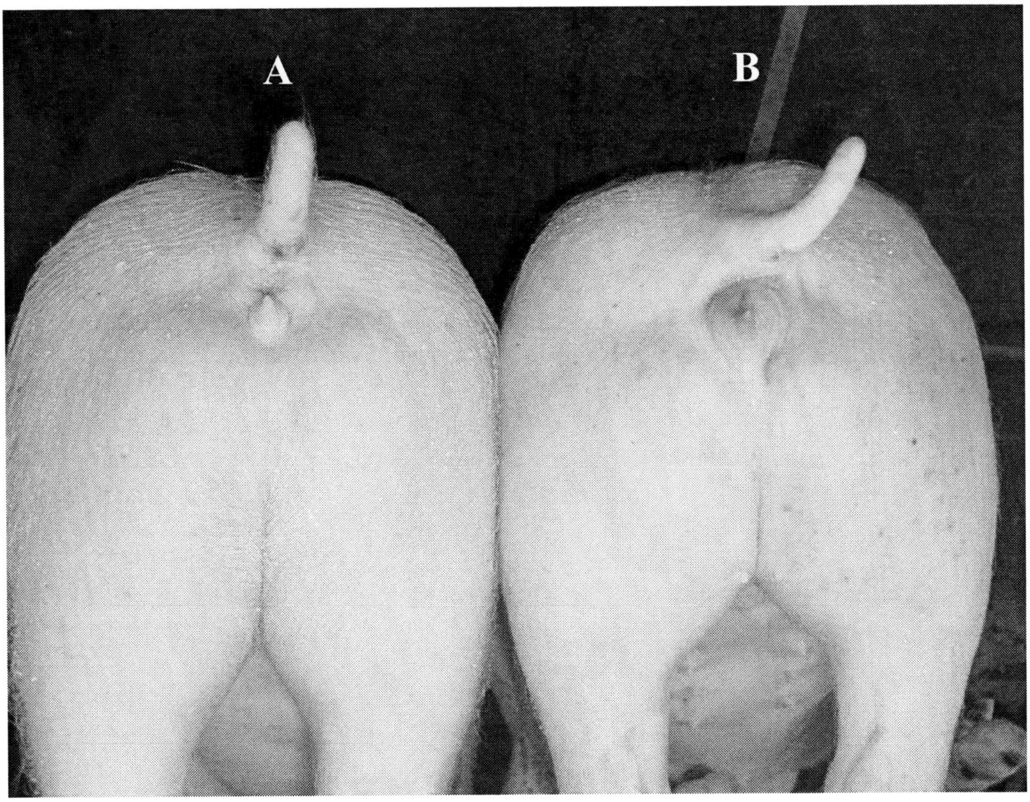

Figure 4. (A) normal; (B) hyperestrogenism syndrome in pigs, medically known as vulvovaginitis picture.

Common signs in pigs poisoned by zearalenone are: infertility, pseudopregnancy (*pseudocyesis*), permanent estrous, reduced number of piglets born, probably due to the reduction on the development of follicles, reabsorption of embryos, malformations and hyperestrogenism in young gilts, that typically occurs after a week of birth females from poisoned matrices [16-18]. The hyperestrogenism is characterized by the resulting reddening and increase the volume of the vulva and early development of the mammary gland. The most critical period of embryo survival seems to be between the seventh and the 10th day of gestation, according to investigation done in poisoned sows. Because estrogens have luteotropism in swine, the probability of abortion in the last two months of pregnancy is low. The denial or reduction in feed intake is often associated with zearalenone poisoning. This occurs because of the unpleasant taste that development and intense fungal contamination confer to feed. In addition, the nutritional value of grains is reduced, particularly energy, which is partially consumed by fungal mass [3, 17].

One of the great losses of toxicosis in pigs by zearalenone is due to abnormalities in estral cycles. Very often, females have all clinical characteristics, but do not demonstrate the positive "reflex of tolerance" in the back pressure test in response to the boar. The indexes of conception in the flock may be reduced by up to 70% and pregnant sows have small litters [17, 18]. Associated with anestro, which is often prolonged, such abnormalities affect significantly the number of piglets born/sow/year and farm productivity. The vitality of piglets within 24 hours after birth is also severely affected. Consequently, greater difficulty to feed occurs, increasing also significantly the number of deaths by starvation and crushing, in addition to increasing the number of stillborn. The number of piglets viable in the first week of life is also affected. Many piglets present opened members syndrome (*splayleg*) and incoordination of the members [3].

Piglets and poisoned sows often have increased and red vulvar lips, and both male and female, have swelling of mammary glands lasting for several days after birth. These clinical signs may come from contact of piglets with zearalenone and/or derived contained in contaminated sow urine, although in most cases, the contamination is still in pregnancy, feeding the sow or, after parturition, through contaminated milk [3, 16-20].

The young boars are generally more affected than adults. The main clinical signs of the toxicological picture in these animals include growth of mammary glands, oedema of the prepuce, getting to dysuria and testicular atrophy, with loss of sperm quality and volume. However, the reduction of libido is also frequently observed, although, apparently, those aspects do not interfere in conception rates when adult boars are used in natural mating scheme [21-22].

The greatest losses, in most cases, are related to animals for breeding, although pigs in growth and termination stages are also affected and the major losses are represented by the reduction in feed intake of the order of 100 to 150 g/day, leading to significant loss of productivity of affected animals at that stage [3].

Several other mycotoxins are produced under the same ideal climatic condition as the production of zearalenone, therefore, with relative frequency. Demonstrations of refusing feed and vomiting are assigned to the instance of concomitant zearalenone and deoxynivalenol toxicosis [3].

Macroscopic Lesions

The primary lesions resulting from poisoning pigs by zearalenone are related to physiological changes in the genital tract and characterized mainly by the accumulation of liquid forming interstitial oedema, cell proliferation and metaplasia of the squamous epithelial cells of the vagina and cervix. The vulva, vagina, cervix and uterus are increased by the combination of cellular oedema with hypertrophy and hyperplasia of the component elements of their structures (Figure 5). Uterotropic effects of zearalenone induce significant increase of the uterus, especially of uterine horns [17].

The ovaries of pre-puberty gilts have increased volume with large number of small follicles, but without evidence luteum corpus formation. Older female have great development of secondary follicules and other follicles show atresia, however, there is no evidence of ovulation, because the ovaries do not have *corpus luteum* in the ovaries of poisoned females. The mammary gland and the teats are oedematous with significant increase in volume [3, 20].

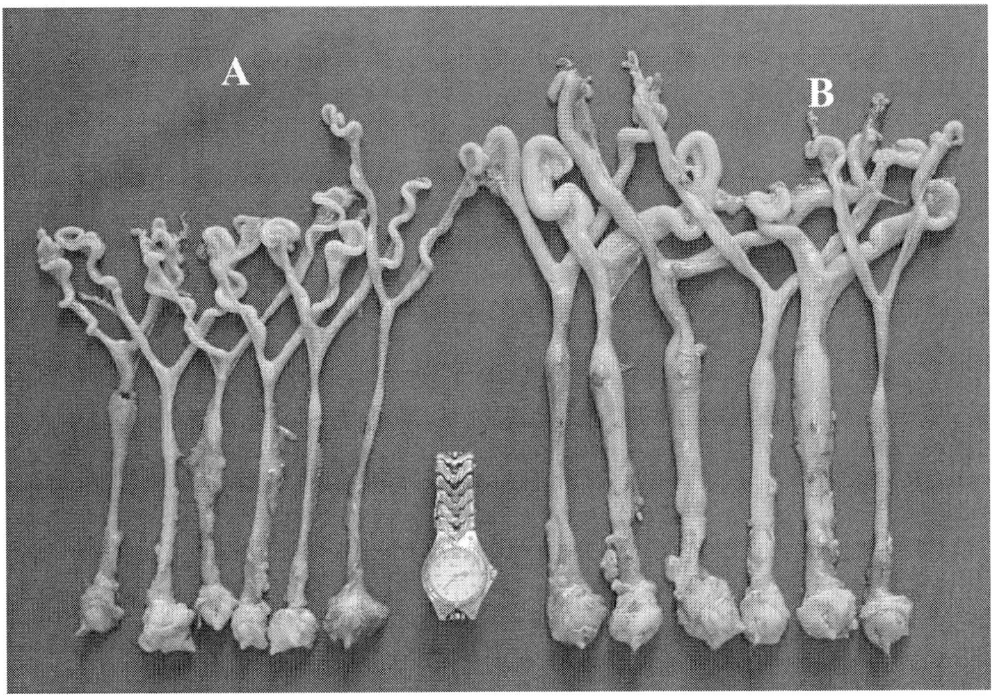

Figure 5: Vulva, vagina, cervix, uterus and ovaries of pigs; (A) normal; (B) intoxicated with zearalenone.

Diagnosis

The clinical characteristics of this toxicosis, which require approximately two weeks to manifesting, involve preventive measures, especially in the previous diagnosis of the contamination of feed. Change of feed is often an indicative factor of the occurrence of infirmity, because usually the first clinical signs appear after 1-3 weeks of use of a new batch. The evaluation of macroscopic characteristics of raw material that makes up the diet alone does not guarantee feed properly analysed and free of zearalenone [2, 3].

Proper monitoring ration constituents, allowing the preventive intervention, is the only way to avoid poisoning pigs by this mycotoxin. The use of anti-mycotoxins additives can be a valid alternative to control [3].

FUMONISINS

Fumonisins form a group of mycotoxins produced by the secondary metabolism of various toxigenic fungi of the genus *Fusarium* and *Alternaria*, being lines of *F. moniliforme* (= *F. verticillioides*) recognized as the largest producers. Nearly two tens of fumonisins are well known. However, only fumonisins B_1 (FB_1), B_2 (FB_2) and B_3 (FB_3) have occurrence and toxicological of significant importance. They occur in various cereals, especially maize, where they have concentrations that usually induce sub-clinical poisoning in several species. The mechanism of action proposed for the toxins is based on the inhibition of the enzyme acetyl transferase, responsible for the processing of sphingolipids. As a result, accumulation of sphinganine and sphingosine in the blood circulation occurs. Equines have the greatest sensitivity to the FB_1 among the domestic species, and the affected animals have essentially clinical signs of nervous disease and lesions of leukoencephalomalacia [23, 24].

Swine have also high sensitivity to fumonisins. In pigs toxin can induce clinical signs and pulmonary oedema. Clinical signs characteristic of acute fumonisin-toxicosis are evidenced by pigs respiratory difficulty and cyanosis resulting from formation of pulmonary oedema [25]. Chronic toxicosis is more frequent and characterized by decreasing the productivity of the herd [26]. In addition to producing characteristics lesions in these species, liver damage are present in all outbreaks of fumonisin-toxicosis. The clinical incidence of pulmonary oedema caused by fumonisin was still too little studied in Brazil [25, 27].

Occurrence of Fumonisins in Feed

Fumonisins are mycotoxins, secondary metabolites of fungi, in particular from the genus *Fusarium*. The largest producer of fumonisin is *F. moniliforme*, but other species are also producing, as *F. proliferatum*, *F. nygamai*, *F. anthophilum*, *F. dlamini* and *F. napiforme*. Recently, it has been demonstrated that fungi of the genus *Alternaria* spp. are also fumonisins producers, however, in smaller quantities. The fungi of the genus *Fusarium* develop and produce larger quantities of fumonisins, preferably in climate conditions of mild temperatures, between 15 and 25°C, and high humidity of the substrate, preferably above 20% or water activity higher than 0.9. By these characteristics, the fungi of the genus *Fusarium* are often called "field fungi", because the largest concentrations of toxins are observed in cereals exposed in the crop to such basic conditions for the fungal development [3, 28, 29].

In the recent evaluations, which involved the last 11 years, carried out on samples from the major grain producer States, LAMIC/UFSM examined fumonisins in 27,857 samples, found average contamination of $FB_1 + FB_2$ of 1,390.0 μg/kg. Maize in this total represented 13,756 samples, with an average contamination 1,239.3 μg/kg, found at 62% of samples of corn. Rations subject to analysis showed contamination in 60.4% of the samples, with an average of contamination near 1,347.0 μg/kg [13].

Pathogeny and Immunosuppressive Effects

Fumonisins have similar molecular structures to sphinganine and sphingosine, responsible for forming a complex in animals (sphingomyelin and glicosphingolipids). Studies using hepatocytes of animals have shown that fumonisins are blocking the formation of sphingolipids through the inhibition of the enzyme N-acyl transferase. Although the mechanism of action is not fully known, a schema that represents the mechanism was proposed based on the similarity of fumonisins' molecules with the sphingolipids [30]. Accumulation of sphinganine and, sometimes, sphingosine occur. In pigs, change in the cardiac dynamics occurs and it was suggested that the formation of pulmonary oedema in pigs could be due to deficiency on the contraction on the left side of the heart and not to the increase on lung capillary permeability [26].

The immunopathogenic potential of fumonisins is feared in swine activity. However, few real importance scale searches of fumonisins in reducing immune capacity of poisoned animals. Thereafter, statistical studies have proved the involvement of FB_1 in the occurrence of reproductive and respiratory syndrome of pigs (Porcine Reproductive and Respiratory Syndrome/PRRS). Researchers concluded that the association of two illnesses is highly significant in 21 herds studied, because from the 12 who had clinical signs of disease, eight had available feed with more than 20 mg FB_1/kg of feed and from the nine herds without the disease, only one was consuming feed with more than 20 mg FB_1/kg of feed. The decrease in concentration of macrophages in lung could be observed from 1 week of toxicosis in piglets fed with feed containing 20 mg of FB_1/kg. Also associated to this low immunity, susceptibility to infection by *Pseudomonas aeruginosa* has been increased. Increased susceptibility to *Escherichia coli* in poisoned piglets has also been proven [2, 3]. Doses of 0.5 mg of FB_1/kg body weight/day, for a period of 7 days were sufficient to provide a significant increase in bacterial spread, when the *Escherichia coli* was administered orally. Although animals do not produce lesions derived from FB_1 poisoning, the researchers concluded that bacterial colonization after 24 hours is significantly greater in the lungs, spleen, kidneys and even greater in digestive organs as ileum, cecum, mesenteric lymph nodes, and colon. They also noted that the poisoned animals with FB_1 had smaller quantities of local inflammatory cytokines [3].

Clinical Signs

The acute toxicosis in pigs by fumonisins induces to a clinical picture with characteristics of pulmonary oedema, which usually occurs between 3 and 5 days after the start of the ingestion of contaminated feed. In such cases, death can occur within a few hours. The chronic poisoning occurs because of the consumption of feed with low concentrations of toxin for a prolonged period, causing liver damage, especially in pigs affected. FB_1 concentrations exceeding 10 mg/kg of feed are considered unsafe for pigs, but pulmonary oedema in natural outbreaks have been seen in pigs fed with contaminated ration from 1 mg FB_1/kg [25-27].

Several researchers found that pigs in reproductive age are more susceptible to fumonisins compared with piglets in growth stage. However, stressful factors as nutritional and management deficiencies may also have highlighted importance in bringing about signs of toxicosis. Females have lower sensitivity than males. However, pregnant poisoned females typically have offspring with lung lesions characteristic of pulmonary oedema. Levels of fumonisins in the milk of poisoned sows with FB_1 contained in ration in concentrations greater than 100 mg/kg of feed were not detected, and the unweaned piglets did not show signs of toxicosis. Abortion in pregnant poisoned by fumonisins generally did not induce complications in subsequent pregnancies [25].

The main clinical signs in chronically poisoned pigs by fumonisins are innespecific and occur as a result of liver lesions, leading to an appearance of bad general state, with decreasing weight gain. Signs can be easily confused with malnutrition, genetic deficiency, inadequate management and/or other diseases which also induce the poor animal clinic appearance. The flock usually shows a significant increment of nonuniformity on the development of animals. Thus the animals reduce feed intake and weight gain, and can also significantly worsen the feed convertion rate. Present bristles with less brightness intensity and erected and animals move with greater frequency in the room without taking feed [3, 25, 27].

Acute toxicosis of pigs by fumonisins is clinically easy to be detected, being characterized by pulmonary oedema signals. More characteristic changes of affected animals include anorexia, lethargy, open-mouthed, increased respiratory frequency, jaundice, and cyanosis on the skin, which are more evident in the ears and nose, sclera and mucous membranes. Excessive drooling, hepatic encephalopathy syndrome, wet stertor to auscultation of the lung and aqueous diarrhoea also occurs. Respiratory difficulties occur normally between 12 and 24 hours after the fall of feed intake highly contaminated and are represented by superficial and abdominal breathing associated with increasing frequency. However, the characteristic signs of pulmonary oedema appear only about 3 to 5 days. Then, more serious signs can be observed, like ataxia of the hindquarters, dog sitting position, decubit pad side, agony for a short period of time and death. Hemoptysis and bleeding faeces can appear in some animals intoxicated. Typically more than 80% of the animals in the same flock are affected, and the recovery can achieve 100% when toxicosis is not very serious and contaminated feed is immediately replaced. On the other hand, all animals intoxicated can die in a few hours if toxicosis is very serious [3].

Macroscopic Lesions

In necropsy of pigs intoxicated with high doses of fumonisins, extensive pulmonary oedema can be observed (Figure 6), which often evolves to hydrothorax, jaundice and darkened and firm liver [25]. Low doses for extended periods of time induce progressive degeneration gravely and hepatic necrosis, interfering in proteic synthesis, weight gain and animal performance. The nefrose of medullary region due to toxicosis by fumonisins is usually an injury associated with hepatosis. Pathological lesions observed in pigs with pulmonary oedema, are the presence of transudate light beige colour or yellow gold in thoracic cavity, characterizing the pulmonary oedema and hydrothorax in more severe poisoning, and quantities exceeding as much as 500 mL of golden-yellow liquid can be found, which starts the process of coagulation when contacting air [2, 25-27]. The lungs of pigs with pulmonary oedema do not collapse when it is removed from the chest, showing increased volume with rounded edges and getting heavier by liquid concentration in lung parenchyma and ducting. Oedema occurs predominantly in the interlobular and interstitial regions. When the accumulation is interlobular, predominance is in the hilo, stretching by the lung, separating lobed with 3-5 mm spaces [25]. The exiting of transudate through pulmonary surface can be even observed, by distended septa and remoteness of the pleura, separating the lung parenchyma. Little liquid is usually found in bronchioles, bronchi and trachea [27].

Few pathological changes can be found in various organs of affected pigs by the pulmonary oedema. However, some researchers report that more flaccid ventricular wall of the heart can occur [26]. The incidence of ulcerative injuries in the esophagus and stomach sometimes associated with liver and esophagical hyperplasic formations can be observed in some animals chronically intoxicated. The increased size of mediastinal and peribronchial lymphonodes in animal intoxicated was found in some affected animals [2].

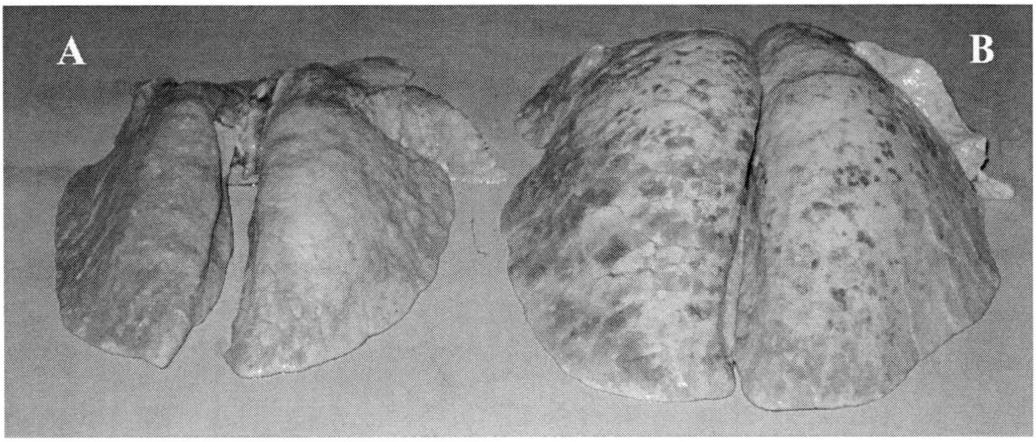

Figure 6. Lungs from pig: (A) normal lung; (B) lung with oedema from pig intoxicated with fumonisins.

Diagnosis and Control

The diagnosis of pulmonary oedema is easily accomplished by observation of clinical signs. However, the definitive diagnosis is based on observation of injuries and the presence of fumonisin on feed analysis.

Toxicosis of pigs by fumonisins has no specific treatment. Clinical signs of acute animal disappear around 3 days after removal of contaminated feed [27].

Prophylaxis consists in adopting measures that inhibit the fungal development on cereals and products used in the diet, and/or adoption of a programme of monitoring the presence of mycotoxin in feed, and avoiding mycotoxin be provided in the diet for a sufficiently long period of time to induce toxicosis [3].

Tricothecenes

The Trichothecenes (TCT) form a group of toxic metabolites chemically similar among them. They are produced by various fungal species of the genus *Fusarium*, *Cephalosporium*, *Myrothecium*, *Stachybotrys* and *Trichoderma*. More than two hundred mycotoxins are part of this group, being subdivided into macrocyclics and non-macrocyclics TCT. The toxic effects of macrocyclic TCT, like baccharins, roridines, satratoxins and verrucarins have not yet been studied in pigs. The non-macrocyclics TCT are divided into type A and type B. The main members of type A comprise the toxin T-2, toxin HT-2 and Diacetoxycirpenol (DAS). The major components of the type B are the Deoxynivalenol (DON or vomitoxin), Fusarenon-X and Nivalenol. The TCT occur in various cereals and by-products, especially when grown in winter whose mild temperatures and high humidity favours the development of fungi producers of these mycotoxins. The toxic action mechanism of TCT is based on inhibition of proteic synthesis and interference with RNA and DNA synthesis, reaching especially cells of high metabolic activity and active multiplication. They lead to the formation of ulcerative lesions in the digestive system and dermatitis, in addition to the refusal of feed intake, which is important for the clinical diagnosis. Therefore, the animals show reduced productive performance and immune system depression. The TCT of more toxicological importance and cost-effective in swine raising are represented in particular by Deoxynivalenol (DON; vomitoxin) and T-2, followed by Diacetoxyscirpenol (DAS) and Nivalenol (NIV) [2, 3].

Occurrence of Trichothecenes in Feed

The occurrence of TCT depends on several factors, such as: strains and its toxigenicity and weather condition including water activity, preferably higher than 0.9. This occurs in particular when harvesting is held under too much rain, cloudy periods and low temperatures ranging between 6 and 24°C (temperate climates). In Southern Brazil, TCT have high incidence on cereals grown in winter, period when fungi have appropriate temperature and humidity for development [3]. Significant concentrations are often found in oats, barley, corn, wheat and hay, while lower levels are usually found in rice, rye and sorghum [2, 13]. Natural contamination levels of DON, T-2 and nivalenol usually reach up to 10 mg/kg. With few exceptions, levels detected range between 15 and 40 mg/kg. The incidence of DON in feeds has occurred very often in levels ranging from 0.1 to 42 mg/kg, with average between 2.4 and 4 mg/kg [13].

Worldwide, DON is the most common contaminant of cereals, accompanied in certain regions by nivalenol (NIV). Other TCT and other *Fusarium* toxins as fumonisins, moniliformin and zearalenone may be concomitantly present in the same cereal batch.

In the last 10 years, on assessments undertaken in LAMIC/UFSM, the incidence of various TCT in feed and raw materials from major grain producers States in Brazil have been studied. The incidence of T-2 was analysed in 13,615 samples. The mean contamination was 9.9 μg/kg in less than 2.2% of positive samples. Nevertheless, the presence of DON was more expressive, being found in 9,705 (37.2%) of 26,047 samples searched with average level of 248 μg/kg, especially being observed in winter cereals, such as wheat and barley [13].

The adverse effects of trichothecenes result from the interaction of toxin or its metabolites directly into the target tissue. Most of the toxic effects of trichothecenes are derived from the inhibition of the primary proteic synthesis, followed by a decrease in production of nucleic acid. The immediate reflexes are recorded in the cell membrane, enzyme activity and immune function. The specific activity, site is the enzyme peptidiltransferase, integrating of ribosomes, it is responsible for elongation and termination of the polypeptidic chain [1, 31].The TCT are also associated with haemorrhages in humans and animals. Prothrombin time is increased significantly, but the primary factor for haemorrhage is by a decreasing in the factor VII of blood coagulation. They cause lymphoid necrosis in the thymus, spleen and lymph nodes. The production of immunoglobulin and serum antibodies is reduced [31].

Pathogeny

The TCTs are one of the most potent inhibitors of protein synthesis of eucaryotic organisms known today. Its inhibitory action is effective in various organisms, including fungi, plants and animals. However, most of the toxic effects of TCT are derived from the inhibition of the primary proteic synthesis, through the inhibition of the enzyme responsible for elongation and polypeptidic chain termination, peptidil-transferase, the ribosomal enzyme of the subregional 60-S unit, presumably blocked by ligation with TCT molecules [1]. However, the primary mechanism of action consists on inhibition of the initiation of the polypeptidic chain, due to affecting the t-RNA and ribosomal sub-units of connection. Immediate reflexes are seen in the function of cells membrane, enzyme activity and immune function. The most affected cells by inhibition of proteic synthesis are those of rapid proliferation, as intestinal, mucous, lymphoid cells, and erytrocytes. The second group of cells very affected belongs to vital organs such as the liver, kidney and pancreas. It was proposed that the greater toxicity of TCT in young or immature animals occurs due to the fact of metabolism speed in these animals be also increased.

Cytotoxicity of TCT was proven in erytrocytes' membrane, and the mechanism of action is not sufficiently understood. It is known, however, that the quantity required to obtain a significant effect is much larger than can usually be detected in the blood circulation of animals suffering by toxicosis of TCT. The main effect of inhibition of production of nucleic acid seems to be to induce, secondarily, the decrease in the production of protein. The RNA seems to be inhibition through a nonspecific secondary mechanism, *i.e.* by decreasing the production of proteins whose synthesis are coordinated. The inhibition of DNA production is linked to the lack of new proteins, recently synthesized to begin the S-phase of DNA replication. In this way, deficiency of DNA is also linked to the decreased production of proteins and vice versa. Due to the damage to DNA, the TCT are recognized powerful as teratogenic and carcinogenic agents, however the carcinogenicity is not so expressive, although products of their biotransformation could exercise a significant effect.

The induction of TCT poisoning resistance, found in mammalian and yeasts cells, is assigned to changes on 60-S ribosome sub-units, reducing affinity to link with all TCT. Another type of resistance is attributed to a recessive gene responsible for the codification of L-3 protein of 60-S ribosome, which is the largest structural protein constituent of the enzyme peptidil-transferase. The mutant protein can increase the resistance of some cell to TCT by up to 100 times.

A second resistance mechanism has been verified in cells that do not have ribosomic change, but have less permeability to toxins [1, 31, 32].

Immunossupressive Effects

The TCT belong to a select group of mycotoxins are of great immunossupressive importance. The inhibitory capacity of toxins over the proteic synthesis and their interaction with the cellular membrane apparently influences on the immune system supressor capacity. In severe acute toxicosis, serious deleterious effects as degeneration and necrosis can be observed in cells of rapid division, as in the intestinal mucosa, spleen, thymus, bone marrow and lymph nodes. The immunoglobulin and serum antibodies production are reduced. Various harmful effects on the immune system cells, on the host's resistance and production of immunoglobulins have already been recorded in several experimental poisonings. Chronic poisoning by TCT on animals induces a significant increase on the susceptibility to various pathogens, such as: *Cryptococcus*; *Candida*; *Listeria*; *Mycobacterium*; *Salmonella* and simple Herpes type 1. The interaction TCT to infectious diseases seems to occur when the resistance mechanism depends on cellular immunity. Thus, the phagocytic capacity is lowered significantly on the occasion of toxicosis by T-2, and may trigger the occurrence of aspergilosis in affected animals. The effect seems to be induced mainly by the decrease of neutrophilic migration, of macrophagic phagocytosis, decrease of rejection of transplanted organs, and delayed hypersensitivity type IV. The decrease of immunity induced by the TCT toxicosis is evidenced by the increased incidence of specific diseases, or those caused by opportunistic pathogens, causing secondary infections. However, intoxicated animal organisms also show an unsatisfactory response to vaccination [2, 3].

Clinical Signs and Lesions

Pigs and other monogastric show great sensitivity to the TCT. NIV and DON have similar, toxigenic potential, inducing the denial of feed and weight loss in pigs intoxicated. A combined level less than 0.4 mg/kg of feed is described as acceptable for pigs, while concentrations exceeding 2.0 mg/kg of feed are always unacceptable [3]. Biological responses to TCT occur within a few minutes after contact, suggesting a rapid absorption in exposed areas, which, under natural circumstances of exposure are, specially, the digestive system and skin. Toxicosis of pigs by trichothecene mycotoxins should consider that other produced by fungi of the genus *Fusarium* as fumonisins, moniliformina and zearalenone may also be present in contaminated feed. The possibility of the presence of multiple trichothecene concurrently on the same feed is also very important [3, 32, 33].

Although the inhibition of proteic synthesis seems to be the largest biochemical effect of TCT, it was not yet determined which proteins are significantly inhibited. However, TCT affect cells in active division, such as the digestive system, skin, lymphoid and erythroid organs. The haemorrhages (observed in intoxicated animals) represented a consequence of ulcerations induced by toxins associated to the decrease in blood clotting factors. Haemorrhagic lesions and clots are frequently observed in the digestive system, which can even induce to bleeding [33].

The complete refusal of feed intake can be observed in pigs where concentrations of DON are greater than 15 mg/kg, however, starting from 1 mg/kg, a decline in dietary intake can be seen. Feed with concentrations between 0.3 mg/kg up to 0.5 mg/kg induces only weight loss, and the refusal of feed might not manifest significant clinical form. Concentrations of 1.3 mg/kg are of economic importance due to the decrease on the performance of animals, and in levels exceeding 2.5 mg/kg, the gain rate decreases by 50%. From 11 mg/kg, animals have weight loss and a complete denial of feed intake. Concentrations exceeding 20 mg/kg of feed drive them vomiting [2, 3].

The most characteristic clinical signs of TCT toxicosis in pigs relate to decrease intake, rejection and vomiting feed, between 15 minutes to approximately one hour after ingestion of contaminated feed, bleeding diarrhoea, reduced feed efficiency, reduction of weight gain, reduction of body weight. Ingestion of 12 μg/kg causes dermatitis, salivation, haemorrhages, abortion and nervous disorders [34]. Macroscopic lesions after necropsy are not always clear, although an increase in the volume of the liver, haemorrhage in lymph nodes, and erosions in the stomach and intestines can be observed in some cases. Characteristic lesions are rarely observed in natural poisoning outbreaks of pigs by trichothecenes, by the fact that the pigs feed intake decrease when toxin is present. The classic signs of intoxicated pigs by TCT are the decreased feed intake and reduced weight gain [3, 32-34].

Figure 7. Vomiting of pig intoxicated with DON.

Deoxynivalenol

Among the different domestic animal species, pigs are the most susceptible to deoxynivalenol (DON) toxicosis. Differences can also be found between the sexes, because males are more susceptible than females. Characteristic clinical signs of this toxicosis in pigs are declining ingestion or complete refusal of feed intake, digestive disorders, vomiting and dccrcasc gain wcight to losc wcight in thc most scrious cascs [2, 3, 32].

Macroscopic lesions of DON toxicosis are rarely observed in sick animals. However, there are reports of animals who expressed changes after a week of toxicosis, like discreet to moderate reddening in the base of the stomach and intestine, associated with an oedematous increase of mesenteric lymph nodes. Sometimes liver weight increase and decrease plasma protein levels, albumin, calcium and phosphorus inorganic can also be observed. Histological lesions consist of vascular congestion of the mucosa of the stomach and intestine. In the stomach, discreet to moderate erosion in the squamous region with lymphocytes and macrophages infiltration and multifocal degeneration of epithelial cells in the base of the organ can still be observed.

Degeneration and necrosis can occur in lymphoid organs, like Peyer's plates and intestinal, mesenteric and bronchial lymph nodes, accompanied by lymphoid depletion. These changes can also sometimes occur in tonsils, thymus and spleen [3, 32].

Toxin T-2

Moderate toxicosis by T-2 includes toxin concentrations usually between 2 and 8 mg/kg of feed. However, young pigs are more resistant and can tolerate up to 10 to 12 mg/kg of feed. However, it is considered that the toxin's secure dose for pigs is below 0.5 mg of toxin per kg of feed. Complete feed refusal is observed consuming feed contaminated with 16 mg/kg. However, significant reductions on feed intake may occur when consuming feed with 3 mg of T2/kg of feed [3].

T-2 induces injuries, especially in the upper regions of the digestive system. However, the consumption of feed with high concentrations of toxin may induce diarrhoeal and perineal injuries, due to remaining toxin in contact still in faeces. This suggests that the toxin has a direct action without metabolic activation. The increased incidence of various types of tumors with increased intake of toxin suggests that the imunossupressive effects or promoters of the toxin could be specially related to effect of starter for tumor process. T-2 is related to the inhibition of proteic synthesis across many systems and its toxic effects are identified primarily active cell division that can explain its teratogenicity [1].

Glucose metabolism is significantly affected in pigs intoxicated by T-2. The hypoglycemia is due to poor absorption of carbohydrates and the increased consumption of glucose. In parallel, the ability to produce glucose by use of amino acids by cortisol is reduced. A reduction on the concentration of fatty acids in intoxicated animal occurs also associated to the decrease in glucose concentration, which is an indication of energy metabolism disorder. The increase in serum activity of enzymes as aspartate aminotransferase starting from concentrations of 0.5 mg of toxin/kg of feed is indicative of liver injury. Pig with toxicosis of concentrations from 2 mg of T-2/kg of feed is associated to significant reduction of lymphocytes, erytrocytes and hemoglobin concentration. The decrease of T-lymphocytes concentration can be found from 0.5 mg toxin/kg feed [1, 33].

Macroscopic lesions on the skin can be observed after a week of intake of poisoned feed containing at least 4 mg of T-2/kg. Dermatitis and crusted formations appear preferably in the skin of the snout, in the upper region of nose, lips' commissures, in the back of the ears and around the prepuce. Similar lesions are also very frequent in the oral mucosa and in the dorsal side of the tongue. However, T-2 also has cardiotoxic properties. Macroscopically, can be observed subendotelial haemorrhages and multifocal white petechiae in the myocardium. Microscopically, multifocal oedema, mononuclear cells infiltration, hyalinization of myofibrils, vacuolization and uncoordinated contractions of stripes of myofibrilas with pyknotic nuclei may occur. Ultrastructurally, areas of swelling and fibrilar disorganization can be observed. Clinically, intoxicated pigs have heart debit, blood pressure, blood circulation reduced in various organs and systems, as well as increased heart rate [3].

Reproductive problems are hardly observed in pigs intoxicated by trichothecene in the level usually found in cereals and feed. Experientially, T-2 cause infertility, abortions, small piglets and decreases the number of piglets per litter.

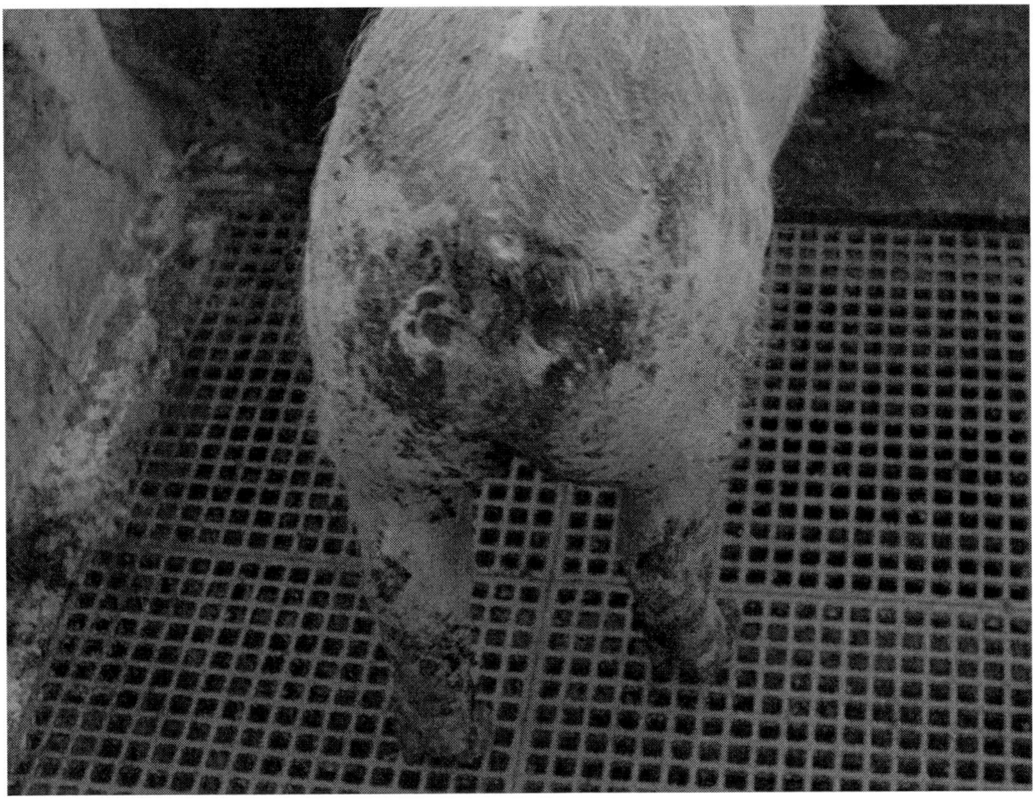

Figure 8. Dermatitis and crusted formations appear on pig intoxicated with T-2 toxin.

Nivalenol and Diacetoxyscirpenol

Pigs intoxicated with nivalenol until 5 mg/kg have not significantly decreased weight gain, clinical signs of feed poisoning or refusal. Lesions due to nivalenol toxicosis rarely occur. However, nephropathy and gastric ulcerations have been already observed in intoxicated pigs with feeds containing 2.5 up to 5 mg/kg. After 3 weeks of toxicosis, the number of cells of the spleen and the concentration of IgG decreases significantly, while a clear increase in concentration of IgA occurs [2, 3].

Diacetoxyscirpenol (DAS) is one of the most toxic among trichothecene mycotoxins. Above of 2 mg/kg of DAS significant toxic effects in pigs always are presents. Decrease feed intake occurs, presumably resulting from the lesion in the central nervous system and due to oral and intestinal lesions induced by toxin. Anorexia is evident when pigs eat feed containing concentrations higher than 10 mg/kg of feed. Sublethal DAS toxicosis results in cell depletion and necrosis of the lymphopoietic organs, multifocal necrosis in the intestinal epithelium and diffuse necrosis in the germ line epithelium, culminating with tubular progressive degeneration in the testicles. Erosive lesions can be observed in the oral cavity of intoxicated pig. After a few days of toxicosis lymphopenia, neutropenia and anaemia occur [3, 31].

Diagnosis and Control

The presumptive diagnosis is based on observation of clinical signs of intoxicated animals through environmental data analysis regarding the harvest and storage of grain used to feed pigs. However, the definitive diagnosis is performed by the analysis of the presence of mycotoxin in feed of intoxicated animal.

For all species chronically affected by TCT, there is an excellent prognosis, since contaminated feed source is removed. Specific therapy for treating intoxicated animals by trichothecene is not yet scientifically proven. The change from contaminated by another free feed of toxin must be the first measure. Symptomatic treatment and support therapy can assist the recovery of affected animals [3].

OCHRATOXIN A

The ochratoxins form a group of seven toxic metabolites produced by fungi of genus *Aspergillus* and *Penicillium* and composed by an isocoumarin structure linked to the amino acid L-β-phenylalanine [1]. Although a number of ochratoxins are known, but only ochratoxin A presents toxicological importance. They occur in a large number of cereals and grains including corn, coffee, beans and peanuts. Pigs have high sensitivity to this mycotoxin, being one of the most toxic mycotoxins for this species. Records about natural toxicosis by ochratoxin A in pigs demonstrated their occurrence in concentrations from 200 μg of toxin per kg of feed eaten. The main organ affected is the kidney. At the molecular level interference in DNA and RNA metabolism occurs through the inhibition of the enzyme t-RNA synthetase. Interference with the m-RNA also occurs, changing gluconeogenesis by inhibition of the enzyme phosphoenolpyruvate. The main clinical signs of ochratoxicosis are related to increased water intake and polyuria. Decrease in feed intake, feed conversion rate and worse weight gain occurs when feeds ingested is contaminated with moderate to high doses of this mycotoxin. In necropsy, or on the occasion of the slaughter of intoxicated animals, kidneys are increased in volume and pale-yellowish or greyish coloured. Histologically, the renal illness is characterized by tubular degeneration, interstitial fibrosis and glomerular degeneration, which reduces the absorption of electrolytes and proteins, increasing urinary excretion throughout osmotic retaining the glomerular filtrate that will be excreted.

In Brazil, among more than 4,000 batches of ration analysed at LAMIC/UFSM, the presence of this toxin was detected in only 3% of the samples, being found in only one time a contamination higher than 744,8 μg/kg [13].

Clinical Signs and Lesions

Toxicosis by ochratoxin A generally occurs during spring or summer. In some farms toxicosis occurs randomLy, and in such cases the illness is usually associated with feed that had serious problems of storage in some stage of conservation. In acute toxicosis, often exceeding 1 mg ochratoxin A /kg feed, pigs have intense thirty and may increase by four times the water consumption. As a result, polyuria occurs as characteristic of the toxicosis. Very often, especially those animals which have difficulty of access to water, animals drink their own urine. These signals can be accompanied by a decline in feed intake, low feed efficiency, diarrhoea and bleeding urine [2, 35].

Doses of 200 μg/kg of feed were sufficient to produce nephropathy in animals, leading to negative results on feed convertion and weight gain. In such cases, clinical signs of ochratoxicosis are quite discreet, showing decrease on feed consumption and weight gain, and polydipsia and polyuria also observed, with lumbar and abdominal depletion, clinical signs that precede the death that arises in a day or two. Mortality can reach 90% in affected flocks. In males, a possible interference in spermatic quality is frequent [2, 3, 35].

Macroscopic Lesions and Clinical Pathology

Although several organs may be altered in swine ochratoxicosis, kidneys are the most affected ones and may present some alteration, like: increase in volume, yellowish or pale colour, and cystic or fibrotic characteristic. When the renal disease is at the last stage, starting from the formation of moderate to severe fibrosis, kidney has ripples or rough surface, because of internal mass reduction [36-38].

Although ochratoxin A affects mainly the kidneys, causing significant increments in serum enzymes activities and blood components such as reductions on protein concentrations, urea, cholesterol, triglycerides, and potassium, the diagnosis of ochratoxicosis is accomplished by analysis of serum ochratoxin A concentration [37]. The quantification of circulating ochratoxin A in the blood associated with clinical signs, especially in endemic regions, are the methods of evaluation of ochratoxin A toxicosis. Changes in the volume and composition of urine are also characteristic of ochratoxin A toxicosis in pigs. In addition to polyuria, increasing the amount of urine eliminated is also significant. The main changes in the composition are related to the increase in concentration of protein and glucose. The presence of blood might indicate serious renal lesions at the nephron level or presence of haemorrhagic tumor processes induced by chronic ochratoxicosis [2, 3].

Effects on the Immunological System

Ochratoxin A is eliminated from the organism in 3 to 4 weeks after the removal of contaminated feed. The difficulty of rapidly removing toxin from the body confers ochratoxin A highly toxic properties to various organs and systems. It has immunossupressive properties in foetuses, young and adult animals. The main effects are related to the decreased phagocytosis and lymphocytic action. Therefore, susceptibility to bacterial infections is significantly increased. Immune response also expected to vaccination is significantly affected, especially in young pigs [3].

Diagnosis and Control

Clinical diagnosis of ochratoxicosis can be performed in endemic areas. Thirty and polyuria suggest the presence of the toxin in the diet. Changing the concentration of enzymes such as gamma-glutamiltransferase and phosphoenolpiruvate also may assist in diagnosis. However, ochratoxin A has peculiar pharmacokinetic properties. Because of the high stability and relatively long half-life of ochratoxin A in the blood, it can be detected in the plasma/serum for a period of 30 days, when ingested with diet in a single dose. Therefore, the quantification of this mycotoxin in the blood plasma has high diagnostic value. It is possible the use of blood from a single animal as a reference of toxin's presence in the flock, as a result of the close correlation between animals that compose the flock that has been fed with the same batch of feed 2, 3].

There is no specific treatment for pigs poisoned by ochratoxins. However, some measures should be immediately implemented to reduce the effects of the toxin in exposed animals. The removal of contaminated ration by feed free of ochratoxin is always the first measure to be implemented. Support treatments can be used to improve the organic conditions of affected animals.

Studies have been developed to reduce the half-life of ochratoxin in the blood circulation intoxicated animal. Testing with phenobarbital administered orally in doses of 80 mg/kg bw/5 days or oral administration of 20 mg 3-metilcolantrena/kg bw/2 days made animals more resistant to ochratoxin A toxicosis, but also increased significantly the LD_{50} values. More detailed studies are needed to increase the efficiency of these products in order to prevent or treat swine ochratoxicosis [3].

CONCLUSION

Mycotoxins represent a great challenge to be overcome in swine production, as they have several deleterious effects on the health and the development of these animals. Because swine feed is extremely dependent on cereals and grains exposed to fungal contamination since the growth of plants, good agricultural practices of cereal cultivation must be implemented with the objective of reducing the contamination of grain destined for pig feed. Then, the grain producers need to be aware about the importance of agricultural practices that seek to reduce contamination of cereals and so also be compensated for the quality of grain, and not only by the volume produced. Prophylactic practices are very importance towards guaranteeing the feed quality on mycotoxins, because only aflatoxins have adsorbent really effective to reduce significantly their toxic effects on animals. Detoxification of other mycotoxins in contaminated feed has not consolidated and proven effectiveness yet, although some products already have demonstrated protector effects. Once contaminated by mycotoxins, feed also has very important losses on nutrient. Therefore, apart from the animals suffering the toxic effects of mycotoxins, they might also be with nutritional deficiencies, because of the diminution of the nutritional value of this feed. In tropical countries, more than two thirds of the raw materials used in pig feed have at least one mycotoxin, and frequently several mycotoxins can be diagnosed in the same feed. That is why a monitoring program of raw materials to be used in animal nutrition is of fundamental importance. A good monitoring program of mycotoxins in raw materials for pig feed must start with a good sample collection program. These samples shall have a good representation on the whole lot. Preferably samples should be collected from the already ground feed and in movement, because it has a better distribution of the particles in the mass and the chances of each particle to be collected is equal for all. So, collective sampling is done and then, mixtures and subsamples are collected, which must be sent to the laboratory to carry out the analyses. Many methodologies may be employed for analysis of mycotoxins. They all have their advantages and disadvantages. It is important, however, that the employed technique is properly validated and reliable. Once in possession of analytical results, it is of fundamental importance that at least the young animals and those destined for breeding purposes are fed with high quality feed and free of mycotoxins.

References

[1] OMS (ORGANIZACIÓN MUNDIAL DE LA SALUD). *Criterios de salud ambiental 11: Mycotoxins*. México: OMS, 1983.

[2] Mallmann, CA; Dilkin, P. Micotoxinas: Clínica e Diagnóstico em Suínos. In: Corrêa, M. N; Lucia, JR.; T; Deschamps, J. C. *Tópicos em Suinocultura II*. Pelotas: Autores: 2003; 1-312.

[3] Mallmann, CA; Dilkin, P. *Micotoxinas e micotoxicoses em suínos*. I Ed. Santa Maria, RS, Brazil: Sociedade Vicente Pallotti; 2007.

[4] Christensen, CM; Nelson, GH. Mycotoxins and micotoxicoses. *Modern veterinary practice,* 1976, 57, 367-371.

[5] Cast, Council for Agricultural Science and Technology. *Micotoxins: Economic and Health Risks*. Report; 1989.

[6] Dilkin, P. Mycotoxicose suína: aspectos preventivos, clínicos e patológicos. *Biológico*, 2002, 64, 187-191.

[7] Cysewsky, SJ; Pier, AC; Engstrom, GW; Richard, JL; Doughterty, RW; Thurstrom, JR. Clinical phathologic features of acute aflatoxicosis of swine. *American journal of veterinary research,* 1968, 29, 1577-1590.

[8] Edds, GT. Acute aflatoxicosis: a review. *Journal of the American Veterinary Medical Association,* 1973, 162, 304-309.

[9] Miller, DM; Crowell, WA; Stuart, BP. Acute aflatoxicosis in swine: Clinical pathology, histopatology, and eletron microscopy. *American journal of veterinary research,* 1982, 43, 273-277.

[10] Mallmann, CA; Santurio, JM; Wentz, I. Aflatoxins – Aspectos clínicos e toxicológicos em suínos. *Ciência Rural*, 1994, 24, 635-643.

[11] Dilkin, P; Mallmann, CA; Almeida, CAA; Corrêa, B. Robotic automated clean-up for detection of fumonisins B1 and B2 in corn and corn-based feed by high-performance liquid chromatography. *Journal of chromatography. A*, 2001. 925, 151-157.

[12] Harvey, RB; Kubena, LF; Phillips, TD; Huff, WE; Corrier, DE. Prevention of aflatoxicosis by addition of hydrated sodium calcium aluminosilicate to diets of growing barrow. *American journal of veterinary research,* 1989, 50, 416-420.

[13] LAMIC/UFSM. Results 2009/08/10. Available from: http://www.LAMIC.ufsm.br/Resultados.

[14]Harvey, RB; Huff, WE; Kubena, LF; Phillips, TD. Progression of aflatoxicosis in growing barrows. *American journal of veterinary research*, 1988, 49, 482-487.

[15]Wilson, DM; Sangster, LT; Bedell, MB. Recognizing the signs of porcine aflatoxicosis. *Veterinary medicine,* 1984, 7, 974-976.

[16]Chang, WM; Kuttz, JH; Mirocha, CJ. Effects of the mycotoxin zearalenone on swine reproduction. *American journal of veterinary research*, 1979, 40, 1260-1267.

[17]Long, GG; Turek, J; Diekman, MA; Scheidt, AB. Effect of zearalenone on days 7 to 10 post-mating blastocyst development and endometrial morphology in sows. *Veterinary pathology*, 1992, 29, 60-67.

[18]Shreeve, BJ; Patterson, DSP; Roberts, BA; Wrathall, AE. Effect of moldy feed containing zearalenone on pregnant sows. *Brazilian Veterinary Journal*, 1978, 134, 421-427.

[19]Edwards, S; Cantley, TC; Rottinghaus, GE; Osweiler, GD; Day, BN. The effects of zearalenone on reproduction in swine. II. The effect on puberty attainment and postweaning rebreeding performance. *Theriogenology*, 1987, 28, 51-58.

[20]Kordic, B; Pribicevic, S; Muntanola-Cvetkovic, M; Nikolic, P; Nikolic, B. Experimental study of the effects of known quantities of zearalenone on swine reproduction. *Journal of environmental pathology, toxicology and oncology*, 1992, 11, 53-55.

[21]Vanyi, A; Szeky, A. Fusariotoxicosis. 6. The effect of F-2 toxin (zearalenone) on the spermatogenesis of male swine. *Magyar állatorvosok lapja*, 1980, 35, 242-246.

[22]Ruhr, LP; Osweiler, GD; Foley, CW. Effect of the estrogenic mycotoxin zearalenone on reproductive potential in the boar. *American journal of veterinary research*, 1983, 44, 483-485.

[23]Mallmann, CA; Santurio, JM; Dilkin, P. Equine leukoecephalomalacia associated with ingestion of corn contaminated with fumonisin B_1. *Journal of Brazilian Society for Microbiology*, 1999, 30, 249-252.

[24]Wilson, TM; Ross, P.; Rice, LG; Osweiler, GD; Nelson, HA; Owens, DL; Plattner, RD; Reggiardo, C; Noon, TH; Pickrell, JW. Fumonisin B_1 levels associated with and epizootic of equine leukoencephalomalacia. *Journal of veterinary diagnostic investigation,* 1990, 2, 213-216.

[25]Dilkin, P; Hassegawa, R; Reis, TA; Mallmann, CA; Corrêa, B. Intoxicação experimental de suínos por fumonisinas. *Ciência Rural*, 2004, 34, 175-81.

[26]Smith, GW; Constable, PD; Eppley, RM; Tumbleson, ME; Gumbrecht, LA; Haschek-Hock, WM. Purified fumonisin B_1 decreases cardiovascular funtion but does not alterpulmonary capillary permeability in swine. *Toxicology Science*, 2000, 56, 240-249.

[27]Dilkin, P; Zorzete, P; Mallmann, CA; Gomes, JDF; Utiyama, LL; Oetting, LL; Corrêa, B. Toxicological effects of chronic low doses of aflatoxin B_1 and fumonisin B_1-containing *Fusarium moniliforme* culture material in weaned piglets. *Food and chemical toxicology*, 2003, 41, 1345-1353.

[28]Dilkin, P; Mallmann, CA; Almeida, CAA; Stefanon, EB; Fontana, FZ; Milbradt, EL. Production of fumonisins by strains of *Fusarium moniliforme* by varying conditions of temperature, moisture and growth period. *Brazilian journal of microbiology*. 2002, 33, 111-118.

[29]Mallmann, CA; Santurio, JM; Almeida, CAA; Dilkin, P. Fumonisin B_1 in cereals and feeds from southern Brazil. *Arquivos do Insituto Biológico*, 2001, 68, 41-45.

[30]Schroeder, JJ; Cranes, HM; Xia, J; Liotta, DC; Merril, AH; Disruption of sphingolipid metabolism and stimulation of DNA synthesis by fumonisin B_1. A molecular mechanism for carcinogenesis associated with *Fusarium moniliforme. Journal of Biological Chemistry,* 1994, 269, 3475-3481.

[31]Ueno Y. Effects of trichothecene mycotoxins on farm animal. Ueno Y. *Trichotecenes, chemical biological and toxicological aspects*. Amsterdam: Elsevier; 1983; 135-146.

[32]Bergsjo, B; Langseth, W; Nafstad, I; Jansen, JH; Larsen, HJS. The effects of natural deoxinivalenol contaminated oats on the clinical condition, blood parameters, performance and carcass composition of growing pigs. *Veterinary research communications,* 1993, 17, 283-294.

[33]Harvey, RB; Kubena, LF; Huff, WE. Effects of tratment of growing swine with aflatoxins and T_2 toxin. *American journal of veterinary research,* 1990, 51, 10, 1688-1693.

[34]Forsyth, DM; Yoshizawa, T; Morooka, N. Emetic and refusal activity of deoxynivalenol to swine. *Applied and environmental microbiology,* 1977, 34, 547-552.

[35]Barnikol, H; Thalmann, A. Clinical observations in the pig in relation to the mycotoxins ochratoxin A and zearalenone. *Tierärztliche Umschau,* 1988, 43, 74-82.

[36]Hult, K; Hokby, E; Sellyey, G; Rutqvist, L; Gatenbeck, S. Ochratoxin A occurrence in slaughter-pigs in Sweden and its use as a tool for feed screening programs. *Journal of environmental pathology, toxicology and oncology,* 1992, 11, 39-40.

[37]Braunberg, RC; Barton, CN; Gantt, OO; Friedman, L. Interaction of citrinin and ochratoxin A. *Natural Toxins,* 1994, 2, 124-131.

[38]Friis, C; Brinn, R; Hald, B. Uptake of ochratoxin A by slices of pig kidney cortex. *Toxicology*, 1988, 52, 209-217.

In: Mycotoxicoses in Animals Economically Important
Editors: Edlyane Gonçalez, et. al. pp. 97-115
ISBN: 978-1-61668-195-1

Chapter 4

MYCOTOXICOSIS IN DOMESTIC FOWL

Carlos A.F. Oliveira[1*], E. Gonçalez[2] and K.R.N. Reddy[3]

[1] Faculdade de Zootecnia e Engenharia de Alimentos,
Universidade de São Paulo, Av. Duque de Caxias Norte, 225,
CEP 13635-900, Pirassununga, SP, Brazil.
[2] Centro de Pesquisa e Desenvolvimento de Sanidade Animal,
Instituto Biológico, São Paulo, SP, Brazil
[3]Agroinnova, Centre of Competence for Innovation in the Agroenvironmental Sector,
University of Torino, Torino, 10095 Grugliasco, Italy

ABSTRACT

This chapter describes the main effects of mycotoxins on domestic fowl, especially those related to long-term ingestion of feed contaminated with low levels of toxins, because acute mycotoxicosis are rare in poultry production. The consequences of chronic mycotoxicosis to poultry production are presented, as well as the adverse effects on performance derived from exposure to cereals and poultry feed contaminated with aflatoxins, fumonisins and ochratoxin A. These toxins are of great importance, not only because of their frequency of occurrence but also due to their toxigenic potential to poultry. Considering that domestic fowls can be adversely affected by levels that can be found under natural conditions, there is the potential for these mycotoxins to have a significant negative economic impact on the poultry industry worldwide. Therefore adequate storage of foodstuffs and animal feed is the main method used to prevent mycotoxin contamination in farms. This is of particular importance to prevent the occurrence of chronic intoxication.

* Corresponding author: C.A.F. Oliveira
Telephone: +55-19-35654173 or +55-19-35654179;
Fax: +55-19-35654284
E-mail address: carlosaf@usp.br

Introduction

During the last decades, poultry production has shown to be an extremely dynamic agricultural sector. Technological progress in genetics, nutrition, animal health and management were responsible for enabling the establishment of a highly efficient and competitive operation all over the world. In highly technified systems, any factor that negatively affects production may lead to huge losses. This is the case of natural contaminants of feeds, such as mycotoxins, which cause considerable losses to poultry breeding.

Mycotoxins are secondary metabolites produced by fungi that occur naturally in foodstuffs. These toxins cause a large variety of toxic effects on vertebrates, including men [1]. Exposure occurs predominantly by the ingestion of contaminated food, mainly cereals used in the preparation of animal feed such as corn, wheat, peanuts and sorghum, as well as other raw materials [2].

Contamination of foodstuffs with toxigenic fungi may occur in different phases of production and processing, from sowing to transportation and storage. Besides, as mycotoxins are generally highly chemically stable, they may remain in the foodstuff even after fungi were removed by usual processing and packaging processes.

Diseases caused by mycotoxins are called mycotoxicosis, and are characterized by diffuse syndromes. However, depending on the type of toxin, lesions predominantly affect a given organ or tissue, such as liver, kidneys, epithelial tissue and Central Nervous System. The simultaneous occurrence of two or more mycotoxins is not rare, and toxic effects on the susceptible organism may be synergistic.

The economic impact of mycotoxins includes mortality of domestic animals, increased costs with veterinary care, and decreased production efficiency. Acute mycotoxicosis is rare in poultry production, and chronic exposure to low levels of mycotoxins is responsible for reduced productivity and increased susceptibility to infectious diseases [3].

In Brazil, several mycotoxins have been identified worldwide in foodstuffs destined to human and animal consumption. Aflatoxins, fumonisins and ochratoxins are of great importance, not only because of their frequency of occurrence but also due to their toxigenic potential to poultry.

Aflatoxins

Aflatoxins are produced by fungi in the genus *Aspergillus*, species *A. flavus* and *A. parasiticus*. These toxins were discovered in 1960, after a toxic outbreak in turkeys in England (Turkey X disease). In this outbreak, thousands of birds died after the intake of Brazilian peanut meal [4].

Nowadays, 18 similar compounds are called aflatoxins. The most important ones in medical / hygienic terms are aflatoxins B_1, B_2, G_1 and G_2 [1]. Their biological activity, however, varies: aflatoxin B_1 (AFB_1), the most frequent one in vegetable substrates, has the greatest toxigenic power, followed by G_1, B_2 and G_2 [4].

Aflatoxins are characterized by their high toxicity. In terms of Animal Health, domestic and laboratory species are sensitive to the acute toxic, mutagenic and carcinogenic effects of these compounds, whose target organ is the liver [5]. Similarly, in terms of Public Health, aflatoxins are involved in the etiology of liver cancer in humans following the ingestion of contaminated foodstuffs [6]. Aflatoxins affect many species, such as turkeys, quails, ducks, geese, as well as larger domestic animals including cattle, lambs and pigs [7].

The genus *Aspergillus* and their toxins are distributed worldwide in foodstuffs and animal feed, but may predominantly be found in tropical and subtropical climates. Plants get contaminated by means of contact with fungal spores in the environment, mainly in the soil, during harvesting and drying. The use of inadequate agricultural practices that increase contact of the products with soil, insect damage to the surface of grains, and inadequate storage in humid and unventilated conditions are pointed out as the main causes that favor contamination and growth of toxigenic fungi [2].

Brazilian regulations determined in 1988 a maximum aflatoxin tolerance level of 50 μg/kg, given by the sum of B_1+B_2+G_1+G_2. Although any raw material to be used directly or as an ingredient in animal feed should comply with this standard [8], aflatoxins are extremely frequent, especially in bird feed, due to the fact the main ingredients (corn, wheat, sorghum) are particularly susceptible to *Aspergillus* growth.

The table 1 shows results of some studies on aflatoxin levels in foodstuffs and animal feed in Brazil. These results indicate that a large percentage of samples show aflatoxin concentrations that may potentially affect poultry productivity.

The particular sensitivity of birds to aflatoxins is related to quick absorption by their gastrointestinal system. After being absorbed, AFB_1 immediately binds to albumin and spreads to the tissues, mainly the liver, where it is biotransformed by the liver microsomal system. Biotransformed derivatives may bind to intracellular constituents such as DNA and RNA, and alter protein synthesis functions [4].

It is important to emphasize that aflatoxin concentration tends to increase throughout the chain of production and commercialization of animal feed. Jones et al. [9] analyzed raw material and feed production, and the same feed stored in the poultry facilities. They observed that mean contamination was equal to 1.2, 6.0 and 8.8 μg/Kg aflatoxin, respectively. The same study showed a strong correlation between the period of storage in the poultry facility and aflatoxin frequency and concentration in the feed. These authors reported that optimal conditions for aflatoxin production in stored feed were 70-89% relative humidity, and temperatures between 19 and 27°C. Fraga et al. [10] found high levels of aflatoxin and ochratoxin A in all stages of poultry feed processing. The evaluation of AFB_1 and fumonisin B_1 (FB_1) was performance in poultry feeds in Brazil, and ranged from 1.2 to 17.5 μg/Kg for AFB_1 [11]. Similar aflatoxin levels were observed in poultry feed in other countries, such as India, Slovakia, Kuwait and Morocco [12-15].

Sensitivity to the toxic effects of aflatoxins is considerably variable among animal species. In terms of commercial poultry, ducks are the most susceptible species, followed by turkeys, geese, pheasants and chickens [16]. Dose-response may be different in individuals of the same species, and may vary according to breed, sex, age, and feed composition, among other factors [1]. In many species, males are more susceptible than females, and in general, young are more susceptible than adult animals [6].

Toxic effects of aflatoxins depend on the dose and exposure time, and these factors will determine the occurrence of acute or chronic intoxications. Acute toxic syndrome is related to the ingestion of foodstuffs with high concentrations of aflatoxins, and the following effects may be observed shortly after ingestion: quick deterioration of general condition, loss of appetite, liver pathology (Figure 1), jaundice, hemorrhage and death [5].

Table 1. Concentrations of aflatoxin, fumonisin and ochratoxin A detected in samples of vegetable products and animal feed traded in Brazil

Type of product	Origin of the samples	Frequency [a] (%)	Mean level (μg/kg)	Reference
Aflatoxins [b]:				
Animal feed	Several states	25.5	50-7,800 [c]	91
Corn	Minas Gerais	18.1	83	92
Animal feed	Several states	10.4	241	93
Corn	Southern Brazil	18.2	79	94
Corn	Southeastern Brazil	8.6	35	94
Corn	Rio Grande do Sul	28.9	1,906	95
Animal feed and corn	Rio Grande do Sul	24.9	4-1,906 [c]	96
Sorghum	Several states	12.8	7-33 [c]	97
Animal feed	Rio de Janeiro	56.9	1-32 [c]	98
Fumonisin B_1:				
Corn	Mato Grosso do Sul	100.0	10,590	99
Corn	Paraná	100.0	2,720-4,790 [c]	99
Corn	São Paulo	76.0	100-6,580 [c]	100
Sorghum	Several states	74.2	110-150 [c]	97
Ochratoxin A:				
Corn	Minas Gerais	0	0	92
Animal feed and corn	Rio Grande do Sul	1.7	14-21 [c]	96
Animal feed	Rio de Janeiro	100.0	1-77 [c]	98

[a] Number of positive samples / total of samples analyzed;
[b] Sum of the four fractions (B_1+B_2+G_1+G_2); [c] Minimal and maximal concentrations.

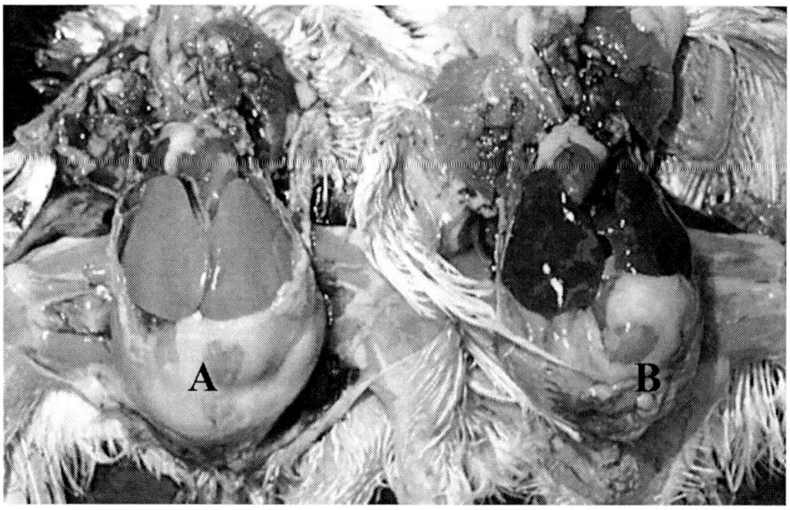

Figure 1. Gross liver pathology of broiler intoxicated with 1.0 mg/kg AFB_1 for 42 days (A) as compared with normal (B).

The most evident clinical sign of chronic aflatoxicosis is a decrease in growth rate of young animals [4] due to a prolonged, continuous or intermittent ingestion of feed contaminated with low levels of aflatoxins. Although this is the main form of intoxication in natural conditions, diagnosis is difficult and economic losses may be considerable [17]. Aflatoxicosis in birds is a highly relevant problem to the broiler industry, because of the economic impact related to mortality, low feed conversion, increased production costs, and negative effects on public health related to human consumption of exposed animals [18].

Aflatoxin B_1 negatively affects chicken (*Gallus domesticus*) growth and the effect is more severe during the development of young animals [18]. Quezada et al. [18] suggest that serum proteins, succinate dehydrogenase (SDH) and glutamate dehydrogenase (GluDH) are sensitive early indicators of this more severe intoxication. Decrease in serum albumin may be used as an early and suitable indicator of the deleterious effects of this mycotoxin in developing chickens.

Giambrone et al. [19] fed broilers for 35 days with diets containing different levels of AFB_1, and observed reduced weight gain and histological changes only in the liver of birds that received more than 500 μg/kg aflatoxin/day. However, in another trial, Giambrone et al. [20] did not observe signs of aflatoxicosis in broilers fed levels up to 800 μg/kg AFB_1 for 5 weeks, although turkeys submitted to the same treatment showed low weight gain and feed conversion, as well as increased morbidity by different causes and higher mortality rates. The authors concluded that levels of up to 66 μg/kg AFB_1 in feed were safe for broilers and turkeys. Results of the study by Kan et al. [21] corroborate this finding, because no differences were observed between broilers offered feed containing 50 to 100 μg/kg AFB_1 and the control group.

Jones et al. [9] analyzed five broiler production companies, and selected 6 growers from each of them. Growers were divided into 3 categories (good, mediocre and poor) according to a productivity indicator (mean market weight X 100 / feed conversion). At the end of the study, the authors observed that frequency of feed contamination in good growers was 18% (mean concentration of 6.13 μg/kg), whereas those classified as mediocre or poor showed contamination frequencies of 22.1 and 31.3% (mean concentrations of 6.5 and 14.0 μg/kg), respectively. Results demonstrated a significant correlation between mortality rates and carcass condemnation in the groups.

In laying hens, the main aflatoxicosis symptoms in experimental conditions included reduced egg production and weight; increased hepatic fat, and changes in serum enzyme levels [4]. Hafez et al. [22] showed atresia in the ovaries of laying hens fed a diet with 8,000 μg/kg AFB_1 for 7 days.

Washburn et al. [23] fed laying hens with 5,000 μg/kg AFB_1 and observed significant decrease in egg weight. This finding was explained by the fact that aflatoxin is primarily metabolized by the liver, which is also the organ responsible for synthesis and transportation of precursors required for yolk production. Similarly, Exarchos et al. [24] and Sudhakar [25] observed adverse effects on egg production when lower concentrations of aflatoxin (600-700 μg/kg) were used for a longer period of time (4-5 weeks).

On the other hand, egg production was not reduced after young laying hens were fed a diet containing 50 μg/kg AFB_1 for 169 days [26]. However, the color of kidneys and livers was slightly lighter when compared with the control group at the end of the trial. Oliveira et al. [27] also showed that there were no deleterious effects on egg production in birds that were given less than 500 μg/kg AFB_1 for a period of 60 days. But these authors reported significant hepatic lesions in birds fed levels over 300 μg/kg. The figure 2 showed bile ducts proliferation of broilers fed with 200 μg/kg AFB_1.

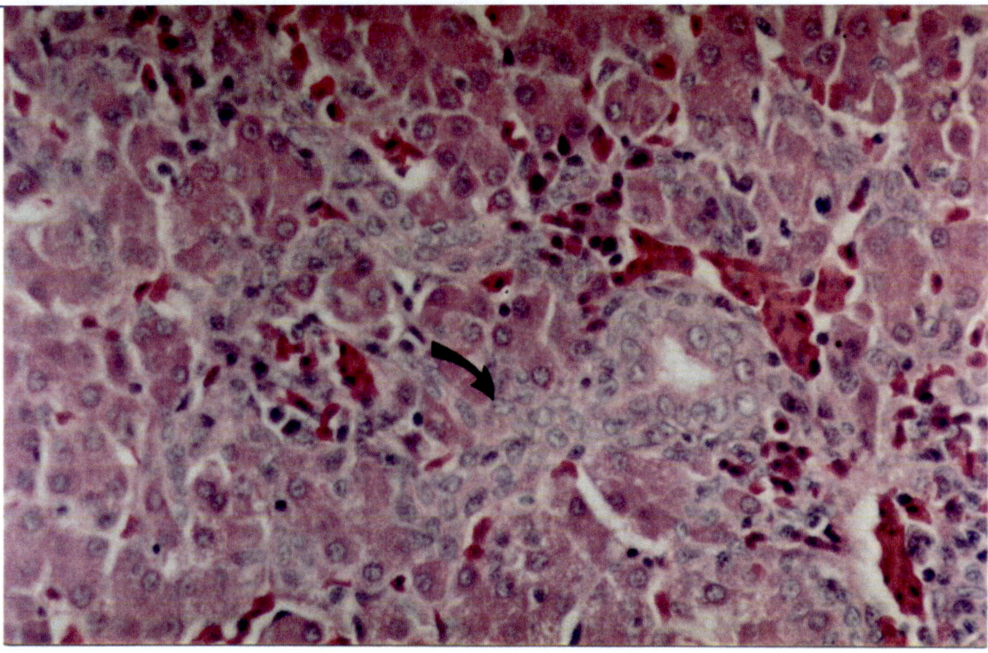

Figure 2. Liver photomicrography of broilers fed with 200 μg/kg AFB1, showing bile ducts proliferation (arrow). Stained with haematoxilin-eosin (Objective 20x).

Broiler breeders [28] and laying quails [29] showed significant decrease in feed consumption and egg production when exposed to levels above 500 μg/kg AFB_1.

Aflatoxins are also capable of affecting the immune system. Immunosuppressing effects demonstrated in domestic birds and other laboratory animals involve aplasia of thymus and bursa of Fabricius, reduction in the number and activity of T cells, decrease in antibody response, suppression of phagocytic activity and reduction in humoral components, such as complement (C4), interferon and immunoglobulins IgG and IgA [17, 30]. All these changes associated with the exposure of animals to feed contaminated with aflatoxins contribute for the occurrence of concomitant infections caused by viral and bacterial agents.

Ghosh et al. [31] observed that 300 μg/kg AFB_1 in broiler feed led to immunosuppression without any apparent clinical effects, and that the ingestion of the toxin may be responsible for morbidity and / or mortality in the flock due to secondary infections. Signs observed in the animals included significant decrease in T cells, albumins and globulins.

In laying hens, the effects of exposure to aflatoxin are a dose-dependent decrease in egg production and egg quality with increased susceptibility to salmonellosis, candidiasis, and coccidiosis [32-35]. Aflatoxin B_1 can have either direct or indirect effects, or both, on the functionality of the gastrointestinal tract of laying hens fed diets containing 0.6, 1.2, or 2.5 mg/kg of AFB_1 for 2 weeks [36].

Significant growth depression, reduced feed consumption and poor feed conversion efficiency, and decreased immune response were observed in broilers fed a diet containing the greatest concentrations of aflatoxins ever studied (1 and 2 mg/kg). Aflatoxins at 2 mg/kg in the diet caused a significant increase in liver relative weight, and a decrease in the relative weight of the bursa of Fabricius [37].

Residues of aflatoxin B_1 were detected in the eggs of hens given 500 μg/kg in the feed, at levels ranging from 0.05 to 0.16 μg/kg (mean of 0.10 μg/kg). Results indicated that the feed-to-egg transmission ratio was approximately 5,000:1, underscoring the importance of controlling aflatoxin levels in laying hen feed [38].

Aflatoxins poisoning in turkeys caused inappetence, reduced spontaneous activity, unsteady gait, recumbence, anemia, and death. In the *post-mortem* examination, body condition was good most of the times, but generalized congestion and edema were observed. Liver and kidney were congested, enlarged, and firm [39].

The presence of aflatoxins in doses equal or greater than 200 ppb affected turkey performance during a period of 1 to 42 days. The effect of those toxins on body weight, feed consumption, relative weight of gizzard and liver, mortality, and total protein and cholesterol levels in serum was dependent on the dose. Turkey poults were very sensitive to aflatoxin poisoning, and inadequate mycotoxicological management may cause important economic losses in turkey production [39].

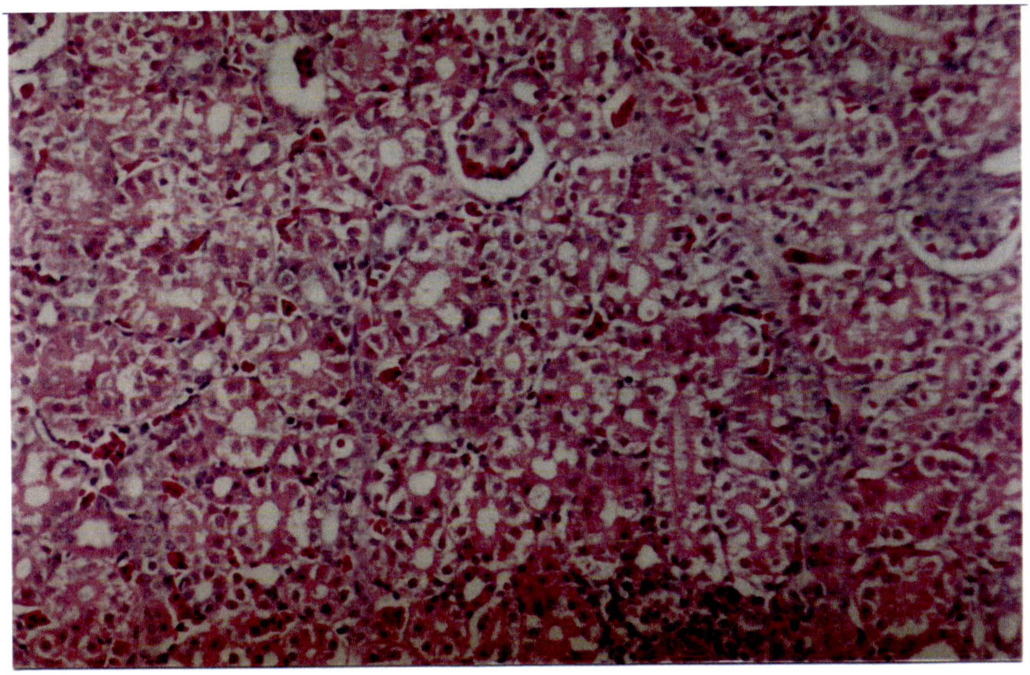

Figure 3. Kidney photomicrography of broilers fed with 200 μg/kg AFB_1 and 200 mg/kg FB1, showing renal tubules with hydropic degeneration. Stained with haematoxilin-eosin (Objective 20x).

Chronic exposure of Japanese quails to diets containing 50 and 100 μg AFB_1/Kg during 168 days adversely affected their performance, especially feed intake, egg weight, and percent eggshell [40]. Addition of 2.5 and 5 mg aflatoxin/kg to quail diets led to significant changes in pancreatic alpha and beta cells. These changes may have a negative effect on carbohydrate metabolism of poultry [41]. Laying quails simultaneous exposed to AFB_1 and fumonisin B_1 in their feed showed primarily additive effects on feed intake, egg and body weight [42], and decreased egg weight, specific gravity and percent eggshell [43].

The effects of prolonged oral administration of aflatoxin B_1 and fumonisin B_1 were evaluated in broilers from 21 to 42 days of age. Results indicated that the levels of FB_1 and AFB_1 studied, alone or in combination, did not cause marked changes in hematological and serological parameters of broilers. However, these mycotoxins may cause important lesions in liver and kidney [44]. Figure 3 showed the renal tubules with hydropic degeneration of broilers fed with 200 μg/kg AFB_1 and 200 mg/kg FB_1.

Broiler chicks were fed different concentrations of AFB_1 and FB_1 (0.50 and 200 μg AFB_1/kg, and/or 0.50 and 200 mg FB_1/kg) from 8 to 41 days of age. All mycotoxin-treated groups showed lower body weight and weight gain, and higher relative heart weight at 41 days. Combination of AFB_1 and FB_1 had primarily additive effects on body weight, liver structure and immunological response at these concentrations [45]. Cockerels fed a control diet contaminated with T-2 and HT-2 toxins (0.31 and 0.26 mg/kg) or with a combination of T-2 and HT-2 toxins (0.32 and 0.25 mg/kg) and aflatoxin B_1 (0.38 mg/kg) for 21 days showed that treatments did not adversely affect production traits and did not exert additive effects on lipid peroxidation and on the glutathione redox system [46].

White Leghorn female chicks treated with different concentrations of aflatoxin B_1 (2.50, 3.13 and 3.91 mg/kg) during 40 weeks had decreased body weight gain and feed intake. Yolk indices and eggshell thickness did not differ significantly, but yolk color was lighter in all aflatoxin-treated groups. Dry matter, crude protein, ether extract, calcium and metabolizable energy retention was adversely affected at the different levels of AFB_1 analyzed when compared with the control group. Pathoanatomic and histopathologic studies showed various adverse changes in liver, kidney, heart, ovaries and bursa of Fabricius in AFB_1-treated groups [47].

FUMONISINS

Fumonisins are a group of mycotoxins recently discovered. Since their isolation in 1988, they have been associated with previously known animal diseases, such as equine leukoencephalomalacia and porcine pulmonary edema [4].

Fumonisins are produced by fungi in the genus *Fusarium,* mainly by the species *Fusarium moniliforme* (synonym *F. verticillioides*) [48]. Other *Fusarium* species are also able to produce these toxins, such as *F. proliferatum* [49], *F. nygamai, F. anthophilum, F. dlamini* and *F. napiforme* [50], as well as fungi in the genus *Alternaria* spp. [51].

Nowadays, 16 molecular structures are called fumonisins [52-53]. However, the predominant toxin produced by *F. moniliforme* lineages is fumonisin B_1 (FB_1) [54]. Only FB_1, B_2 (FB_2) and B_3 (FB_3) have been detected when fumonisin production occurred in natural conditions [55].

Studies with rodent hepatocytes have shown that fumonisins block sphingolipid production. This evidence corroborates the hypothesis that the interruption of sphingolipid production and the consequent accumulation of sphinganine and sphingosine are the main toxic mechanisms of fumonisin acute toxicity and carcinogenicity [54]. Sphingolipids are important in the maintenance of cell membrane integrity, as well as in the regulation of cell surface receptors, ion pumps and other vital systems for cell function and survival [4].

The genus *Fusarium* is worldwide distributed and found in soil and on the surface of plants. Contamination of grain and cereals may occur both in the field and during storage. In natural conditions, this genus grows well in corn. Difficulty in harvesting at the correct maturation stage and high humidity at storage are ideal conditions for toxin production [4].

Due to the predominance of tropical and subtropical climates in Brazil, fumonisin incidence is elevated and some cases show high contamination levels. However, there is no specific regulation that sets safe concentrations of fumonisins in human and animal foodstuffs.

Although fumonisins are highly toxic for horses and pigs, most domestic birds are highly resistant to them [4]. In a study involving broilers fed a diet with 300 mg/kg FB_1 during the first 2 weeks of life, Brown et al. [56] observed diarrhea, 19% weight gain reduction, 30% increase in liver weight, besides histological changes, such as multifocal hepatic necrosis, biliary hyperplasia, muscle necrosis and rachitism.

Kubena et al. [57] observed few deleterious effects on the performance and health of adult laying hens given feed containing 100-200 mg/kg FB_1 and 50-100 mg/kg moniliformin for a prolonged period of time (420 days). On the other hand, Prathapkumar et al. [58] studied natural mycotoxicosis outbreaks in laying poultry, and reported dark and viscous feces, 20% reduction in egg production and 10% mortality. Flocks showed evident improvement after they stopped receiving contaminated feed. Analysis of the feed showed contamination levels of 8.5 mg/kg FB_1 and 100 μg/kg AFB_1, evidencing the additive effects of the two toxins.

Another example of the interaction between these mycotoxins comes from the study by Javed et al. [59] in which different concentrations of FB_1 (61-646 mg/kg), FB_2 (14-98 mg/kg) and moniliformin (M) (66-367 mg/kg), alone or in combination, were offered to broilers of different ages. All groups showed evident clinical signs of intoxication, as well as decreased weight gain and increased mortality rates. Toxic effects were exacerbated by the interaction of FB_1 and M.

Weibking et al. [60] analyzed broilers given feed with different concentrations of FB_1 and observed liver lesions and decrease in weight gain in the groups receiving 225 and 450 mg/kg FB_1, respectively. Compared with the control group, animals that received 75 mg/kg showed only reduced sphingolipid biosynthesis, indicating that although clinical signs may not be evident, and this concentration may be toxic for birds.

Inmunosuppresion is also pointed out as one of the main toxic effects of FB_1. Li et al. [61] did not observe decreases in weight gain or increases in feed conversion in broilers fed 200 mg/kg FB_1. However, they reported decreased humoral immunity and lymphocyte suppression.

Fusarium moniliforme-contaminated culture material has been shown to induce toxic responses when fed to chickens, ducklings, and turkey poults. Toxicity signs included increased mortality, reduced size of the bursa of Fabricius, thymus and spleen, decreased body weight gain, myocardial degeneration and hemorrhage, changes in hemostatic mechanisms, and necrosis of hepatocytes [59, 62-66].

Most of the reports on toxic effects of FB_1 in avian species involve chickens and turkeys. High levels of this toxin fed to these species are associated with reduced body weight and feed intake, diarrhea, poor performance, and alterations in hematological and biochemical parameters, with increased activity of the enzymes alanine transaminase and aspartate transaminase [67]. The latest European recommendations suggest maximum fumonisin levels of 60 mg/kg in corn and 20 mg/kg in poultry feed [68].

Fumonisins significantly increased total serum protein and albumin concentrations on 14 and 21 days post-feeding (DPF), serum calcium and cholesterol levels from 14 DPF onward, and creatinine from 21 DPF onward [67]. This study revealed that the addition of *F. verticillioides* (synonym *F. moniliforme*) culture material supplied 300 ppm of FB_1/kg of diet, a highly toxic concentration to quail chicks, resulting in heavy mortality, decreased growth rate, and significant alterations in hematological and biochemical parameters.

When performance is considered, 50 mg FB_1/kg were detrimental to turkeys but not to broilers fed to market age. Liver sphingolipid results indicated that 25 and 50 mg FB_1/kg increased the sphinganine/sphingosine (Sa/So) ratio, but no significant toxic effect was observed. The altered Sa/So ratio with no concurrent disease suggested that this ratio is an extremely sensitive biomarker for exposure to FB_1 but may not be a good indicator of clinically significant disease in poultry [69].

No impact was observed on feed consumption and growth of turkeys exposed to different concentrations of fumonisins (0, 5, 10, and 20 mg of fumonisin B_1 + fumonisin B_2/kg of feed) over a period of 9 weeks. Weight of tissues and markers of liver damage were not altered, either. However, sphingolipid metabolism was clearly disrupted at levels of 10 and 20 mg [68].

Studies using *F. moniliforme*-contaminated culture material with known concentrations of fumonisin B_1 suggest that fumonisins are toxic to chickens and turkeys at concentrations above 100 mg/kg of feed [60, 70]. Broilers and turkeys are relatively resistant to FB_1 toxicity but exhibited poor performance, increased organ weights, and increased Sa/So ratio. When exposed to high levels of FB_1, turkeys also showed hepatocellular hyperplasia [71] and broilers, multifocal hepatic necrosis [60]. Chronic effects of FB_1 have been studied in turkey poults fed 75 mg FB_1/kg of diet for 18 weeks [72], and in laying hens fed 100 or 200 mg FB_1/kg for 420 days [57]. Turkey poults showed decreased body weight gain and enlarged livers [72]. However, in laying hens, 100 and 200 mg FB_1/kg did not affect body weight gain (at 420 days) or relative liver weight (at 112 days) [57].

As observed in other species, FB_1 is weakly absorbed in turkeys. Mean residence time (MRT) of FB_1 was also relatively high after oral administration to turkeys, which is in agreement with the presence of FB_1 in their liver and kidney demonstrated after 8 hours of ingestion of feed contaminated with 20 mg FB_1 + FB_2/kg [73].

Fumonisin B_1 is toxic to chicken macrophages and decreased their functional activity [74]. Other trials have shown that fumonisins are toxic to chicken embryos, but had no effect on the development of embryos that survived the critical period of 7 days after exposure [75].

Mild to moderate periportal extramedullary hematopoiesis and mild focal hepatic necrosis were observed in chicks fed FB_1 (100 or 200 mg/kg). FB_1 and moniliformin, alone or in combination, adversely affected chick performance and health at dietary concentrations of 100 or 200 mg/kg [76].

The major gross lesions found in affected male chicks fed diets containing fuminisin B_1 and moniliformin included ascites, hydropericardium, hepatopathy, nephropathy, cardiomyopathy, pneumonitis, gizzard ulceration, and enlarged bursa of Fabricius filled with caseous material. Histopathologic and ultrastrutural changes were observed in liver, kidneys, lungs, heart and pancreas [77].

Exposure to FB_1 at concentrations ≥ 50 mg/kg adversely affected quail performance, causing a reduction in feed intake and lower weight gain. Egg production, weight, and eggshell weight were also affected [78]. Interestingly, eggshell de-pigmentation was observed which became more marked with increasing concentrations of FB_1, thus suggesting a dose-dependent relationship (Figure 4). Consequences of this decrease in eggshell pigmentation are difficult to assess at this time. Consumers may associate decreased pigmentation with poor bird health and hence lower quality eggs. Decreased pigmentation may also suggest decreased nutrient content of eggs with negative effects on fertility and hatchability of eggs used to produce laying quail.

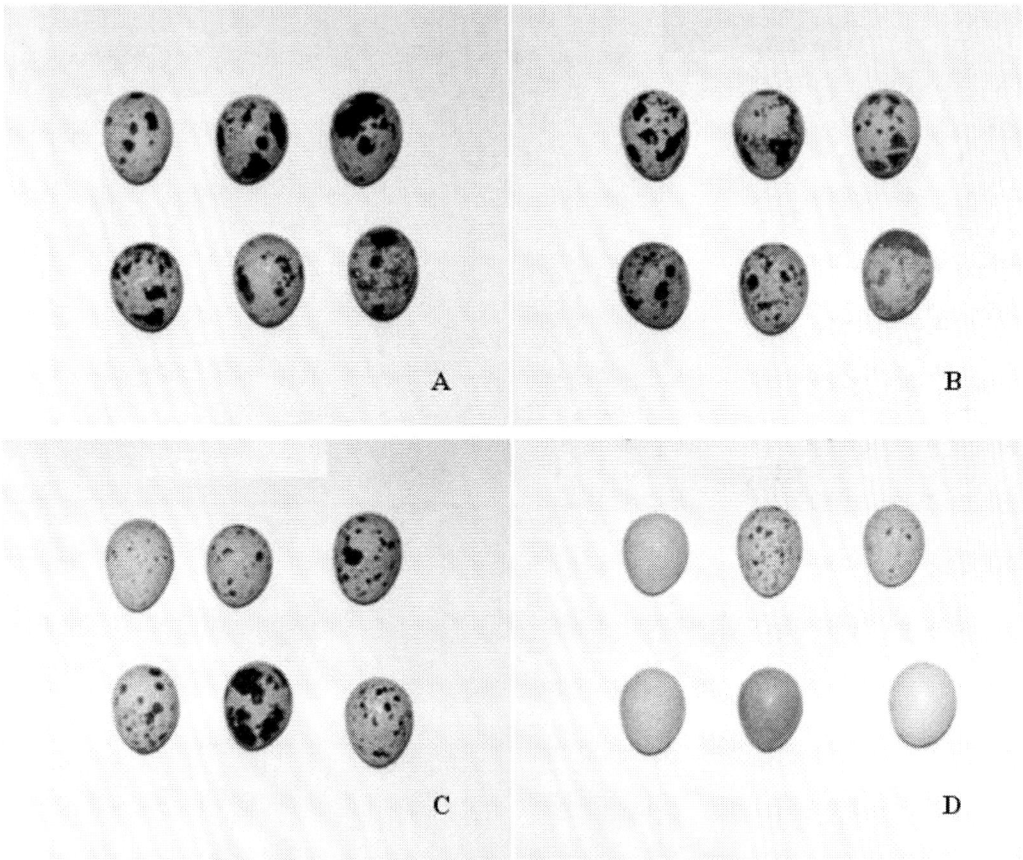

Figure 4. De-pigmentation of eggs from laying quail fed increasing concentrations of dietary FB_1. (A) 0 (control), (B) 10 mg/kg, (C) 50 mg/kg, (D) 250 mg/kg.

OCHRATOXIN A

Ochratoxins were first isolated from the species *Aspergillus ochraceus* (previously known as *A. alutaceus*), these toxins may be produced by several species in the genera *Aspergillus* and *Penicillium* [4]. The ochratoxin group has 7 components, but only ochratoxin A (OA) has been found worldwide as a natural contaminant of grains. The kidneys are the target organs of OA, which interferes in the synthesis of macromolecules in cells of the renal parenchyma, including DNA, RNA and proteins. Besides, it affects renal carbohydrate metabolism, damaging the epithelium of the proximal tubules and causing osmotic diuresis by means of decreased absorption of electrolytes and increased water excretion [4].

Most frequently, OA contaminates grains and cereals such as corn, barley, wheat and rye. Ochratoxin production occurs at room temperature between 4 - 37° C, relative humidity between 18.5 – 40.4% and water activity between 0.83 – 0.90, depending on the toxigenic lineage. Although safe levels of OA in foodstuffs destined to human and animal consumption have not yet been determined in Brazil, other countries have established acceptable thresholds, such as Uruguay (50 μg/kg) and France (5 μg/kg). There are few surveys on the occurrence of OA in foodstuffs and animal feed in Brazil, but the results of some studies are summarized in Table 1. It should be pointed out, however, that given the size of the country, different regions have different climates. Southeastern Brazil, for instance, has optimum conditions for the production of OA by *A. ochraceus* lineages, whereas the southern region has ideal conditions for the genus *Penicillium*. Therefore, studies on the occurrence of this toxin are essential in order to more precisely estimate the level of exposure of human and animal populations.

Figueroa et al. [79] analyzed 50 samples of a commercial brand of poultry feed in Venezuela, and showed that 94% of the samples were contaminated by OA at levels ranging from 2,558 to 31,978 μg/kg, and that in 42% of them levels ranged from 10 up to 20 μg/kg.

OA causes a series of adverse effects in most domestic animals. In pigs, it is the causative agent of mycotoxic porcine nephropathy. Similarly, in humans, it is recognized as the etiological agent of Balkan endemic nephropathy. In poultry, OA is considered to be the most potent mycotoxin, showing high lethality rates. LD_{50} for broilers is around 2.1 mg/ kg of body weight [80], whereas LD_{50} for AFB_1 is 6.8 mg/kg [81].

Hamilton et al. [82] reported the natural occurrence of mycotoxicosis outbreaks and observed the presence of OA in animal feed, at levels ranging from 0.2 to 16.0 mg/kg. These authors reported a mortality rate of 59% and 20% decrease in feed consumption in turkey breeding facilities; nephrotoxicity and secondary airsacculitis were also observed. Outbreaks involving laying hens were characterized by decreased egg production, low quality of eggshell and nephropathy. In broilers, these authors observed low growth rate and feed efficiency, little pigmentation of the carcasses and nephropathy.

Huff et al. [83] and Gentles et al. [88] reported that broilers given more than 2.0 mg/kg OA in their feed showed decreased weight gain, relative increase in kidney and liver weight, besides decreased serum levels of total proteins, albumins, globulins and cholesterol, and increased creatinine and uric acid.

Mraz et al. [85] also reported a reduction in the weight of broilers offered 0.85 mg/kg OA, although no macroscopic changes were observable in their organs. Singh et al. [86] studied broilers fed diets containing 0.5 to 2.0 mg/kg OA, and observed a reduction in the number of lymphocytes T and total lymphocytes, as well as a decrease in albumin and globulin concentration, and reduced relative weight of some organs, such as the bursa of Fabricius, spleen and thymus. Prior et al. [87] reported that laying hens fed diets containing more than 0.5 mg/kg OA showed reduced egg production and feed consumption. Other unknown effects of OA intoxication may lead to different problems in the next steps of poultry production.

Warren et al. [88] observed that broilers fed diets containing 2.0 mg/kg OA showed significant reduction in the breaking strength of the large intestines, besides increase relative weight of the organ. This finding is important when considering the possible rupture of intestines during poultry processing, which would lead to fecal contamination of the carcasses and to their condemnation in the slaughterhouse.

Combination of OA and T-2 toxin caused impaired immune function in chicks even at combined concentrations as low as 0.25 mg/kg OA and 0.5 mg/kg T-2 toxin [89].

Male broilers fed diets containing 400 and 800 ppb OA showed significantly decrease in body and thymus weight, feed consumption, feed conversion ratio and thyroxin concentration ($P<0.05$). OA-treated groups developed anemia manifested by a significant decrease in red cell counts (37%), packed cell volume percentage and hemoglobin concentration. White cell counts, humoral immune response and cell-mediated immunity were also significantly reduced [90].

The greatest concentrations of OA (2 and 4 mg/kg) ever given to broilers produced significant growth rate depression, reduced feed consumption, and poor feed conversion efficiency and immune response [37].

Conclusion

Climate characteristics of Brazil are ideal for the development of toxigenic fungi. Besides, inadequate sowing, harvesting, drying, transportation and storage of agricultural products are still common in some regions of the country. Therefore, the adoption of agricultural practices that prevent the contamination and development of fungi is essential to assure the high quality of raw material to be used in poultry feed production.

Adequate storage of foodstuffs and animal feed is the main method used to prevent mycotoxin contamination in farms. This is of particular importance to prevent the occurrence of chronic intoxication, once prolonged intake of low levels of aflatoxins and fumonisins may lead to several problems related to productive performance in poultry.

References

[1] Coulombe, RA. Aflatoxins. In: Sharma, RP; Salunkhe, DK (editors.) *Mycotoxins and phytoalexins*. Boca Raton: CRC Press, 1991, 103-143.

[2] Chu, FS. Mycotoxins: food contamination, mechanism, carcinogenic potencial and preventive measures. *Mutation Research*, 1991, 259, 291-306.

[3] Hussein, H S; Brasel, JM.. Toxicity, metabolism, and impact of mycotoxins on humans and animals. *Toxicology*, 2001, 167,101–134.

[4] Leeson, S.; Diaz, G.J.; Summers, J.D. (1995). *Poultry metabolic disorders and mycotoxins*. Guelph: University Books.

[5] Osweiler, GD. Mycotoxins and livestock: What role do fungal toxins play in illness and production losses? *Veterinary Medicine*, 1990, 85, 89-94.

[6] McLean, M.; Dutton, M.F. Cellular interacions and metabolism of aflatoxin: an update. *Pharmacology & Therapeutics*, 1995, 65, 163-192.

[7] Fernandez, A; Ramos, JJ; Sanz, T; Fernandez, DL. Alterations in the performance, haematology and clinical biochemistry of growing lambs fed with aflatoxin in the diet. *Journal Applied in Toxicology*, 1996. 16, 85–91.

[8] Brasil. Ministério da Agricultura. Portaria MA/SNAD/SFA nº 07, de 09/11/88. *Diário Oficial da União*, Brasília, 9 nov. 1988. Seção I, p.21968.

[9] Jones, FT; Hagler, WH; Hamilton, PB. Association of low levels of aflatoxin in feed with productivity losses in commercial broiler operations. *Poultry Science*, 1982, 61, 861-868.

[10] Fraga, ME; Curvello, F; Gatti, MJ; Cavaglieri, LR; Dalcero, AM; da Rocha Rosa, CA. Potential aflatoxin and ochratoxin a production by *Aspergillus* species in poultry feed processing. *Veterinary Research Communications*, 2007, 31, 343-353.

[11] Oliveira, GR; Ribeiro, JM; Fraga, ME; Cavaglieri, LR; Direito, GM; Keller, KM; Dalcero, AM; Rosa, CA. Mycobiota in poultry feeds and natural occurrence of aflatoxins, fumonisins and zearalenone in the Rio de Janeiro State, Brazil. *Mycopathologia*, 2006, 162, 355-362.

[12] Thirumala-Devi, K; Mayo, MA; Reddy, G; Reddy, DV. Occurrence of aflatoxins and ochratoxin A in Indian poultry feeds. *Journal Food Protection*, 2002, 65, 1338-1340.

[13] Labuda, R; Parich, A; Vekiru, E; Tancinová, D. Incidence of fumonisins, moniliformin and Fusarium species in poultry feed mixtures from Slovakia. *Annals of Agricultural and Environmental Medicine*, 2005, 12, 81-86.

[14] Mu, B; Al-Mutairi, M; Beg, KR; Al-Mazeedi, HM; Ali, LN; Saeed, T. Mycotoxins in poultry feed in Kuwait. *Archives of Environmental Contamination and Toxicology*, 2006, 50, 594-602.

[15] Zinedine, A; Juan, C; Soriano, JM; Moltó, JC; Idrissi, L; Mañes, J. Limited survey for the occurrence of aflatoxins in cereals and poultry feeds from Rabat, Morocco. *International Journal Food Microbiology*, 2007, 115, 124-127.

[16] Muller, RD; Carlson, CW; Semeniuk, G; Harshfield, GS. The response of chicks, ducklings, goslings, pheasants and poults to graded levels of aflatoxin. *Poultry Science*, 1970, 49, 1346-1350.

[17] Pier, AC. Major biological consequences of aflatoxicosis in animal production. *Journal of Animal Science*, 1992, 70, 3964-3967.

[18] Quezada, T; Cuéllar, H; Jaramillo-Juárez, F; Valdivia, AG; Reyes, JL. Effects of aflatoxin B(1) on the liver and kidney of broiler chickens during development. *Comparative Biochemistry and Physiology - Part C: Toxicology & Pharmacology*, 2000, 125, 265-272.

[19] Giambrone, JJ; Diener, UL; Davis, ND; Panangala, VS; Hoerr, FJ. Effects of purified aflatoxin on broiler chickens. *Poultry Scence*, 1985a, 64, 852-858.

[20] Giambrone, JJ; Diener, UL; Davis, ND; Panangala, VS; Hoerr, FJ. Effects of aflatoxin on young turkeys and broiler chickens. *Poultry Science*, 1985b, 64, 1678-1684.

[21] Kan, CA; Rump, R; Kosutzky, J. Low level exposure of broilers and laying hens to aflatoxin B_1 from naturally contaminated corn. *Archiv fuer Gefluegelkunde*, 1989,53, 204-206.

[22] Hafez, AH; Megalla, SE; Abdel-Fattah, HM; Kamel, YY II. Effects of aflatoxin on ovaries and testicles in mature domestic fowls. *Mycopathologia*, 1982, 77, 137-139.

[23] Washburn, KW; Wyatt, RD; Potts, PL; Lanze, GM. Effects and mechanism of aflatoxin variation in shell strength. *Poultry Science*, 1985, 64, 302-1305.

[24] Exarchos, CC; Gentry, RF. Effect of aflatoxin B_1 on egg production. *Avian Disease*, 1982, 26, 191-195.

[25] Sudhakar, BV. Effect of aflatoxins on egg production and its quality. *Poultry Advances*, 1990, 23, 43-46.

[26] Micco, C; Miraglia, M; Onori, R; Brera, C; Mantovani, AL; Ioppolo, A; Stasolla, D. Long-term administration of low doses of mycotoxins to poultry. 1. Residues of aflatoxin B_1 and its metabolites in broilers and laying hens. *Food Additives and Contaminants*, 1988, 5, 303-308.

[27] Oliveira, CAF; Reis, TA; Albuquerque, R; Guerra, JL; Correa, B. Hepatic lesions in laying hens chronically exposed to rations containing different levels of aflatoxin B_1. *Arquivos do Instistuto Biologico*, 1999, 66, 39-43.

[28] Muthiah, J; Reddy, PR; Chandran, NDJ. Effect of graded levels of aflatoxin B_1 and the effect of direct fed microbials (DFM) on egg production in egg type breeders. *Indian Veterinary Journal*, 1998, 75, 231-233.

[29] Johri, TS; Rashmi, A; Sadagopan, VR. Effect of low dietary levels of aflatoxin on laying quails (*Coturnix coturnix japonica*) and their response to dietary modifications. *Indian Journal of Animal Science*, 1990, 60, 355-359.

[30] Pestka, JJ; Bondy, GS. Alteration of immune function following dietary mycotoxin exposure. *Canadian Journal Physiology and Pharmacology*, 1990, 68, 1009-1016.

[31] Ghosh, RC; Chauhan, HVS; Roy, S. Immunosupression in broilers under experimental aflatoxicosis. *British Veterinary Journal*, 1990, 146, 457-462.

[32] Pier, A.; Richard, J.; Cysewski, S. Implications of mycotoxins in animal disease. *Journal of the American Veterinary Medical Association*, 1980, 8, 719–724.

[33] Wyatt, R. D. (1991). Poultry. In J. E. Smith; R. S. Henderson (editors). *Mycotoxins and Animal Foods*, pp. 553–605. Boca Raton, FL: CRC Press.

[34] Celik, IO; Demet, HH; Donmez, H; Ogˇuz, V; Boydak, M. Determination of phagocytic and candidacildal activities of peritoneal macrophages isolated from chickens fed with aflatoxin adsorbing agent, polyvinylpolypyrrolidone. *Journal of Veterinary Science*, 1996, 12, 145–151.

[35] Kececi, TH; Ogˇuz, V; Kurtogˇ, L; Demet, O. Effects of polyvinylpolypyrrolidone synthetic zeolite and bentonite on serum biochemical and haematological characters of broiler chickens during aflatoxicosis. *British Poultry Science*, 1998, 39, 452–458.

[36] Applegate, TJ; Schatzmayr, G; Pricket, K; Troche, C; Jiang, Z. Effect of aflatoxin culturc on intcstinal function and nutricnt loss in laying hcns. *Poultry Science*, 2009, 88, 1235-41.

[37] Verma, J; Johri, TS; Swain, BK; Ameena, S. Effect of graded levels of aflatoxin, ochratoxin and their combinations on the performance and immune response of broilers. *British Poultry Science*, 2004, 45, 512-518.

[38] Oliveira, CA F; Kobashigawa, E; Reis, TA; Mestieri, L.; Albuquerque, R.; Correa, B. Aflatoxin B1 residues in eggs of laying hens fed a diet containing different levels of the mycotoxin. *Food Additives and Contaminants*, 2000, 17, 459–462.

[39] Rauber, RH; Dilkin, P; Giacomini, LZ; Araújo de Almeida, CA; Mallmann, CA. Performance of turkey poults fed different doses of aflatoxins in the diet. *Poultry Science*, 2007, 86, 1620-1624.

[40] Oliveira, CA F; Rosmaninho, JF; Butkeraitis, P; Correa, B; Reis, TA; Guerra, JL; Albuquerque, R; Moro, MEG. Effect of low levels of dietary aflatoxin B1 on laying Japanese Quail. *Poultry Science*, 2002, 81, 976–980.

[41] Simşek, N; Ergün, L; Ergün, E; Alabay, B; Ozen, A. Ultrastructure of pancreatic alpha and beta cells in young quails (Coturnix coturnix japonica) fed aflatoxin. *Deutsche Tierarztliche Wochenschrifte*, 2007, 114, 465-469.

[42] Ogido, R; Oliveira, CA; Ledoux DR; Rottinghaus, GE; Corrêa, B; Butkeraitis, P.; Reis, TA; Gonçalez, E.; Albuquerque, R. Effects of prolonged administration of aflatoxin B1 and fumonisin B1 in laying Japanese quail. *Poultry Science*, 2004, 83, 1953-1958.

[43] Oliveira, CA; Ogido, R; Ledoux DR; Rottinghaus, GE; Corrêa, B; Reis, TA; Gonçalez, E. The quality of eggs from japanese quail, *Coturnix japonica*, fed rations contaning aflatoxin B1 and fumonisin B1. *Journal of Poultry Science*, 2007, 44, 29-33.

[44] Del Bianchi, M; Oliveira, CA; Albuquerque, R; Guerra, JL; Correa, B. Effects of prolonged oral administration of aflatoxin B1 and fumonisin B1 in broiler chickens. *Poultry Science*, 2005, 84, 1835-1840.

[45] Tessari EN, Oliveira CA, Cardoso AL, Ledoux DR, Rottinghaus GE Effects of aflatoxin B1 and fumonisin B1 on body weight, antibody titres and histology of broiler chicks. *British Poultry Science*, 2006, 47, 357-364.

[46] Pál, L; Dublecz, K; Weber, M; Balogh, K; Erdélyi, M; Szigeti, G; Mézes, M. Effect of combined treatment with aflatoxin B1 and T-2 toxin and metabolites on some production traits and lipid peroxide status parameters of broiler chickens. *Acta Veterinaria Hungarica*, 2009, 57, 75-84.

[47] Pandey, I; Chauhan, SS. Studies on production performance and toxin residues in tissues and eggs of layer chickens fed on diets with various concentrations of aflatoxin AFB1. *British Poultry Science*, 2007, 48, 713-723.

[48] Shephard, GS; Thiel, PG; Sydenham, EW. Initial studies on the toxicokinetics of fumonisin B_1 in rats. *Food Chemistry and Toxicology*, 1992, 30, 277-279.

[49] Ross, PF; Nelson, PE; Richard, ID; Osweiller, GD; Rice, LG; Plattner, RD; Wilson, TM. Production of fumonisins by *Fusarium moniliforme* and *F. proliferatum* isolates associated with equine leukoencephalomalacia and pulmonary edema syndrome in swine. *Applied Environmental Microbiology*, 1990, 56, 3225-3226.

[50] Nelson, PE. Taxonomy and biology of *Fusarim moniliforme*. *Mycopathologia*, 1992, 117, 29-36.

[51] Chen, J; Mirocha, CJ; Xie, W; Hogge, L; Olson, D. Production of the mycotoxin fumonisin B_1 by *Alternaria alternata f. sp. lycopersici*. *Applied Environmental in Microbiology*, 1992, 58, 3928-3931.

[52] Musser, SM; Plattner, RD. Fumonisin composition in cultures of *Fusarim moniliforme, Fusarium proliferatum* and *Fusarium nygami*. *Journal Agriculture and Food Chemistry*, 1997, 45, 1169-1173.

[53] Ah-seo, J; Won Lee, Y. Natural occurrence of the C series of fumonisins in moldy corn. *Applied Environmental in Microbiology*, 1999, 65, 1331-1334.

[54] Norred, WP. Fumonisins - mycotoxins produced by *Fusarium moniliforme*. *Journal Toxicology and Environmental Health*, 1993, 38, 309-328.

[55] Henry, MH; Wyatt, RD. A review of fumonisin production by *Fusarium moniliforme* and fumonisin toxicoses in animals. *Applied Poultry Research*, 1993, 2, 188-192.

[56] Brown, TP; Rottinghaus, GE; Williams, ME. Fumonisins mycotoxicosis in broilers: performance and pathology. *Avian Disease*, 1992, 36, 450-454.

[57] Kubena, LF; Harvey, RB; Buckley, SA; Bailey, RH; Rottinghaus, GE. Effects of long-term feeding of diets containing moniliformin, supplied by *Fusarium fujikuroi* culture material, and fumonisin, supplied by *Fusarium moniliforme* culture material, to laying hens. *Poultry Science*, 1999, 78, 1499-1505.

[58] Prathapkumar, SH; Rao, VS; Paramkishan, RJ; Bhat, RV. Disease outbreak in laying hens arising from the consumption of fumonisin-contaminated food. *British Poultry. Science*, 1997, 38, 475-479.

[59] Javed, T; Bennett, GA; Richard, JL; Dombrink-Kurtzman, MA; Cote, LM; Buck, WB Mortality in broiler chicks on feed amended with *Fusarium proliferatum* culture material or with purified fumonisin B_1 and moniliformin. *Mycopathologia*, 1993, 123, 171-184.

[60] Weibking, TS; Ledoux, DR; Bermudez, AJ; Turk, JR; Rottinghaus, GE. Effects of feeding *Fusarium moniliforme* culture material, containing known levels of fumonisin B_1, on the young broiler chick. P*oultry Science*, 1993, 72, 456-466.

[61] Li, YC; Ledoux, DR; Bermudez, AJ; Fritsche, KL; Rottinghaus, GE. Effects of fumonisin B_1 on selected immune responses in broiler chicks. *Poultry Science*, 1999, 78, 1275-1282.

[62] Jeschke, N; Nelson, PE; Marasas, WFO. Toxicity to ducklings of *Fusarium moniliforme* isolated from corn intended for use in poultry feed. *Poultry Science*, 1987, 66,1619–1623.

[63] Engelhardt, JA; Carlton, WW; Tuite, JF. Toxicity of *Fusarium moniliforme* var. *subglutinans* for chicks, ducklings and turkey poults. *Avian Disease*, 1989, 33, 357–360.

[64] Marijanovic, DR; Holt, P; Norred, WP; Bacon, CW; Voss, KA; Stancel, PC. Immunosuppressive effects of *Fusarium moniliforme* corn cultures in chicken. *Poultry Science*, 1991, 70, 1895–1901.

[65] Weibking, TS; Ledoux, DR; Brown, TB; Rottinghaus, GE. Fumonisin toxicity in turkeys. *Journal of Veterinary Diagnostic Investigation*, 1993b, 5, 75–83.

[66] Espada, Y; Gopegui, RR; Cuadradas, C; Cabanes, FJ. Fumonisin mycotoxicoses in broilers: Plasma protein and coagulation modifications. *Avian Disease*, 1997, 41, 73–79.

[67] Arsani, RK; Katoch, RC; Gupta, VK; Deshmukh, S; Jindal, N; Ledoux, DR; Rottinghaus, GE; Singh, SP. Effects of Feeding *Fusarium verticillioides* (Formerly *Fusarium moniliforme*) Culture Material Containing Known Levels of Fumonisin B1 in Japanese Quail (*Coturnix coturnix japonica*). *Poultry Science*, 2006, 685; 1129–1135.

[68] Tardieu, D; Bailly, JD; Skiba, F; Métayer, JP; Grosjean, F; Guerre, P. Chronic toxicity of fumonisins in turkeys. *Poultry Science*, 2007, 86, 1887-1893.

[69] Broomhead, JN; Ledoux, DR; Bermudez, AJ; Rottinghaus, GE. Chronic effects of fumonisin B1 in broilers and turkeys fed dietary treatments to market age. *Poultry Science*, 2002, 81, 56-61.

[70] Ledoux, DR; Weibking, TS; Rottinghaus, GE. Effects of *Fusarium moniliforme* culture material containing known levels of fumonisin B1 on turkey poults. *Poultry Science*, 1992, 71(Suppl. 1), 162.

[71] Weibking, TS; Ledoux, DR; Bermudez, AJ; Rottinghaus, GE. Individual and combined effects of feeding *Fusarium moniliforme* culture material, containing known levels of fumonisin B1, and aflatoxin B1 in the young turkey poult. *Poultry Science*, 1994, 73, 1517–1525.

[72] Bermudez, AJ; Ledoux, DR; Turk, JR; Rottinghaus, GE. The chronic effects of *Fusarium moniliforme* culture material, containing known levels of fumonisin B1, in turkeys. Avian Disease, 1996, 40, 231–235.

[73] Tardieu, D; Bailly, JD; Skiba, F; Grosjean, F; Guerre, P. Toxicokinetics of fumonisin B1 in turkey poults and tissue persistence after exposure to a diet containing the maximum European tolerance for fumonisins in avian feeds. *Food Chemistry and Toxicology*, 2008, 46, 3213-3218.

[74] Qureshi, MA; Hagler Jr., WM. Effects of fumonisin-B1 exposure on chicken macrophage functions *in vitro*. *Poultry Science*, 1992, 71, 104-112.

[75] Henry, MH; Wyatt, RD. The Toxicity of Fumonisin B1, B2, and B3, Individually and in combination, in chicken embryos. *Poultry Science*, 2001, 80, 401–407.

[76] Ledoux, DR; Broomhead, JN; Bermudez, AJ; Rottinghaus, GE. Individual and combined effects of the *Fusarium* mycotoxins fumonisin B1 and moniliformin in broiler chicks. *Avian Disease*, 2003, 47, 1368-1375.

[77] Javed, T; Bunte, RM; Dombrink-Kurtzman, MA; Richard, JL; Bennett, GA; Côté, LM; Buck, WB. Comparative pathologic changes in broiler chicks on feed amended with *Fusarium proliferatum* culture material or purified fumonisin B1 and moniliformin. *Mycopathologia*, 2005, 159, 553-564.

[78] Butkeraitis, P; Oliveira, CA; Ledoux, DR; Ogido, R; Albuquerque, R; Rosmaninho, JF; Rottinghaus, GE. Effect of dietary fumonisin B1 on laying Japanese quail. *British Poultry Science*, 2004, 45, 798-801.

[79] Figueroa, S; Centeno, S; Calvo, MA; Rengel, A; Adelantado, C. Mycobiota and concentration of ochratoxin A in concentrated poultry feed from Venezuela. *Pakistan Journal of Biological Science*, 2009, 12, 589-594.

[80] Huff, WE; Wyatt, TL; Tucker, TL; Hamilton, PB. Ochratoxicosis in the broiler chicken. *Poultry Science*, 1974, 53, 1585-1591.

[81] Smith, JW; Hamilton, PB. Aflatoxicosis in the broiler chickens. *Poultry Science*, 1970, 49, 207-215.

[82] Hamilton, PB; Huff, WE; Harris, JR; Wyatt, RD. Natural ocurrences of ochratoxicosis in poultry. *Poultry Science*, 1982, 61, 1832-1841.

[83] Huff, WE; Kubena, LR; Harvey, RB. Progression of ochratoxicosis in broiler chickens. *Poultry Science*, 1988, 67, 1139-1146.

[84] Gentles, A; Smith, EE; Kubena, LF; Duffus, E; Johnson, P; Thompson, J; Harvey, RB; Edrington, TS. Toxicological evaluations of cyclopiazonic acid and ochratoxin A in broilers. *Poultry Science*, 1999, 78, 1380-1384.

[85] Mraz, A; Kosutzky, J. Clinical effects and morphological changes after administration of low doses of ochratoxin A to broiler chicks. *Veterinary Medicine*, 1992, 37, 237-242.

[86] Singh, GS; Chauhan, HV; Jha, GJ; Singh, KK Immunosupression due to chronic ochratoxicosis in broiler chicks. *Journal of Comparative Pathology*, 1990, 03, 399-410.

[87] Prior, MG; Sisodia, CS. Ochratoxicosis in White Leghorn hens. *Poultry Science*, 1978, 57, 619-623.

[88] Warren, MF; Hamilton, PB. Intestinal fragility during ochratoxicosis and aflatoxicosis in broiler chickens. *Applied and Environmental Microbiology*, 1980, 40, 641-645.

[89] Wang, GH; Xue, CY; Chen, F; Ma, YL; Zhang, XB; Bi, YZ; Cao, YC. Effects of combinations of ochratoxin A and T-2 toxin on immune function of yellow-feathered broiler chickens. *Poultry Science,* 2009, 88, 504-10.

[90] Elaroussi, MA; Mohamed, FR; El Barkouky, EM; Atta, AM; Abdou, AM; Hatab, MH. Experimental ochratoxicosis in broiler chickens. *Avian Pathology*, 2006, 35, 263-269.

[91] Sabino, M. Variações de níveis de aflatoxina B1 em alimentos e rações animais no período de 1971 a 1979. *Revista do Instituto Adolfo Lutz*, 1980, 40, 153-158.

[92] Sabino, M; Prado, G; Colen, G. Ocorrência de aflatoxinas, ocratoxina A e zearalenona em milho de Minas Gerais. Parte I. *Revista do Instuto Adolfo Lutz*, 1986, 46, 65-71.

[93] Sabino, M; Lamardo, LCA; Inomata, EI; Ichikawa, AH; Giannattasio, CMP. Ocorrência de aflatoxina B1 em produtos alimentícios e rações animais, consumidos no Estado de São Paulo e em várias regiões do Brasil, no período de 1980 a 1987. *Revista do Instituto Adolfo Lutz*, 1988, 48, 81-85.

[94] Sabino, M; Prado, G; Inomata, EI; Pedroso, MO; Garcia, RV. Natural occurence of aflatoxins in maize in Brasil. Part II. *Food Additives and Contaminants*, 1989, 6, 327-331.

[95] Santurio, JM; Baldissera, MA; Almeida, CAA; Ahmad, SHE; Pranke, DHL; Heinrichs, CM; Zanandrea, S. Aflatoxinas, ocratoxina A e zearalenona em grãos e rações destinadas ao consumo animal no sul do Brasil. In: ENCONTRO NACIONAL DE MICOTOXINAS, 7., 1992, São Paulo, SP. *Anais*. São Paulo, SP: 1992. p.14.

[96] Baldissera, MA; Santurio, JM; Canto, SH; Pranke, PH; Almeida, CAA; Schimidt, C. Aflatoxinas, ocratoxina e zearalenona em alimentos para consumo animal no sul do Brasil - Parte II. *Revista do Instituto Adolfo Lutz*, 1993, 53, 5-10.

[97] Silva, JB; Pozzi, CR; Mallozzi, MAB; Ortega, EM; Correa, B. Mycoflora and occurrence of aflatoxin B1 and fumonisin B1 in stored brazilian sorghum. In: *International Iupac Symposium On Mycotoxins And Phycotoxins*, 10, 2000, Guarujá, SP. *Abstract Book*. Guarujá, SP: 2000. p. 145.

[98] Ribeiro, JMM; Rosa, CAR; Curvello, FA; Fraga, ME. Toxigenic mycoflora and mycotoxins (aflatoxins and ochratoxin A) in poultry feed in Rio de Janeiro, Brazil. In: *International Iupac Symposium On Mycotoxins And Phycotoxins*, 10, 2000, Guarujá, SP. *Abstract Book*. Guarujá, SP: 2000, p. 133.

[99] Hirooka, EY; Yamaguchi, MM; Aoyama, S. The natural occurrence of fumonisins in Brazilian corn kernels. *Food Additives and Contaminants*, 1996, 13, 173-183.

[100] Camargos, SM; Soares, LMV; Sawazaki, E; Bolonhezi, D; Castro, JL; Bortolleto, N. Fumonisins in corn cultivars grown during the 94/95 season in the state of São Paulo, Brazil. In: INTERNATIONAL IUPAC SYMPOSIUM ON MYCOTOXINS AND PHYCOTOXINS, 10, 2000, Guarujá, SP. *Abstract Book*. Guarujá, SP. p.142.

In: Mycotoxicoses in Animals Economically Important
Editors: Edlyane Gonçalez, et. al. pp. 117-129
ISBN: 978-1-61668-195-1

Chapter 5

INTERACTIVE EFFECTS BETWEEN MYCOTOXINS IN LIVESTOCKS

Carlos A.F. Oliveira[1] and B. Corrêa[2]

[1] Faculdade de Zooctenia e Engenharia de Alimentos da Universidade de São Paulo (FZEA – USP) Av. Duque de Caxias Norte, 225, CEP 13635-900, Pirassununga, SP, Brazil

[2] Instituto de Ciências Biomédicas da Universidade de São Paulo (ICB – USP) Av. Prof. Lineu Prestes, 1347, CEP 05508-000, São Paulo, SP, Brazil

ABSTRACT

The scientific literature is replete with information on the effects of individual mycotoxins in various livestock species. The occurrence of multiple mycotoxin contamination in feedstuffs has become a greater concern to the animal production, because of the possibility of interaction effects of different mycotoxins. In this chapter, the main interaction effects of mycotoxins on domestic animals are presented, especially those related to long-term ingestion of feed contaminated with low levels of toxins. The toxicity of some individual mycotoxins can be enhanced in a synergistic, additive or antagonistic manner when they occur as co-contaminants and are consumed by different animal models.

INTRODUCTION

Mycotoxins are organic compounds of low molecular weight, produced as toxic secondary metabolites by various species of different fungal genders, with various chemical structures occurring in a variety of substrates, including agricultural products [1].

---

1 Telephone: + 55 19 35654173
E-mail: carlosaf@usp.br
2 Corresponding author: B. Corrêa
Telephone: +55-11-30917295; Fax: +55-11-30917354
E-mail address: correabe@usp.br

Toxigenic fungi can contaminate food in the various stages of production and benefiting, since planting to transportation and storage. As a remark, mycotoxins present, in general, great chemical stability, what enables its persistency in food even after fungi remotion through usual processes of industrialization and packaging.

Diseases caused by mycotoxins are denominated mycotoxicoses, which are characterized by diffuse syndromes, however, presenting injuries in certain organs like liver, kidney, epithelial tissue and central nervous system, depending on the kind of toxin. It is also possible the simultaneous occurrence of two or more mycotoxins, which may lead to the potentiation of the toxic effects over the susceptible organism.

Mycotoxicoses are distributed in three categories: *acute primary mycotoxicoses*, occurring when, considering sensibility, it is consumed in moderate-to-high quantities of mycotoxins, causing a specific acute clinical status; *chronic primary mycotoxicoses*, that occur under consumption of moderate-to-low quantities, causing decrease of reproduction efficiency, weight gain and growth rate; these clinical signs may be mistaken with other pathologies or even to nutritional deficiency; *secondary mycotoxicoses*, result from the action of low levels of the toxin, unable to cause mycotoxicoses, however, predisposing to infectious diseases due to immunosuppression [2].

The main three groups of mycotoxins and toxin-producing fungi were distributed by Clevstrom [3] as follows:

a) Aflatoxin, produced by fungi of the *Aspergillus* genus, mainly *A. flavus* and *A. parasiticus*;
b) *Fusarium* toxins, produced by species of the *Fusarium* genus, which main mycotoxins are fumonisins, trichothecenes, moniliformins and zearalenone;
c) Ochratoxins, produced by *Aspergillus alutaceus* (*A. ochraceus*) and species of *Penicillium* genus.

Other mycotoxin-producing genera include *Claviceps, Alternaria, Pithomyces, Myrothecium, Stachybotrys* and *Phoma*. Various toxins have been identified in human and animal feed, worldwide. However, it is worth mention the importance of the aflatoxins, fumonisins, trichothecenes, zearalenone and ochratoxins, not only for their usual occurrence but also for the high toxigenic potential demonstrated by them in domestic animals.

All these fungi as well as the mycotoxins may grow well under a given set of conditions like: food susceptibility, fungus producing colonization, appropriate temperature, food humidity, relative air humidity during food storage, fungus capacity to produce mycotoxins, presence of insects and mites. Among these aspects, the most influents in contamination are food, humidity and temperature [4].

Toxigenic fungi inhabit the most various substrates like soil, water, vegetables, human and animals. From their natural habitat, fungi spread themselves in nature, through means like the air (the most important mean), water, insects, mites, human and animals. Under favorable conditions and in contact with certain substrates (food, water, soil, etc) fungi grow unsusceptible to produce toxins [5].

Individual Biological Activity of Mycotoxins

Aflatoxins

Aflatoxins are produced by *Aspergillus* fungi, species *A. flavus* and *A. parasiticus*, developed in 1960, after causing a toxic outbreak in turkeys in England (Turkey-X-disease). In this outbreak, millions of birds died after the consumption of peanut meal in the feed, originated from Brazil [6].

There are more than 20 types of aflatoxins molecules and their isolated derivatives, however, the main studied types remains to be B_1, B_2, G_1 and G_2 [7]. The chemical structure of the aflatoxins is really similar, considering that they are simple chemical compounds of low molecular weight, presenting a cumarinic central core connected to a bi-furanoid structure, as shown in figure 1 [8]. Aflatoxins B present ciclopentane ring in the molecule, while aflatoxins G present lactone ring.

Aflatoxins, as well as other heterocycles compounds, are fluorescent substances with proper characteristics. Aflatoxin B_1 (AFB_1) as well as aflatoxin B_2 (AFB_2) presents blue fluorescence, while aflatoxin G_1 (AFG_1) and G_2 (AFG_2) present green-yellow fluorescence under ultraviolet light [7].

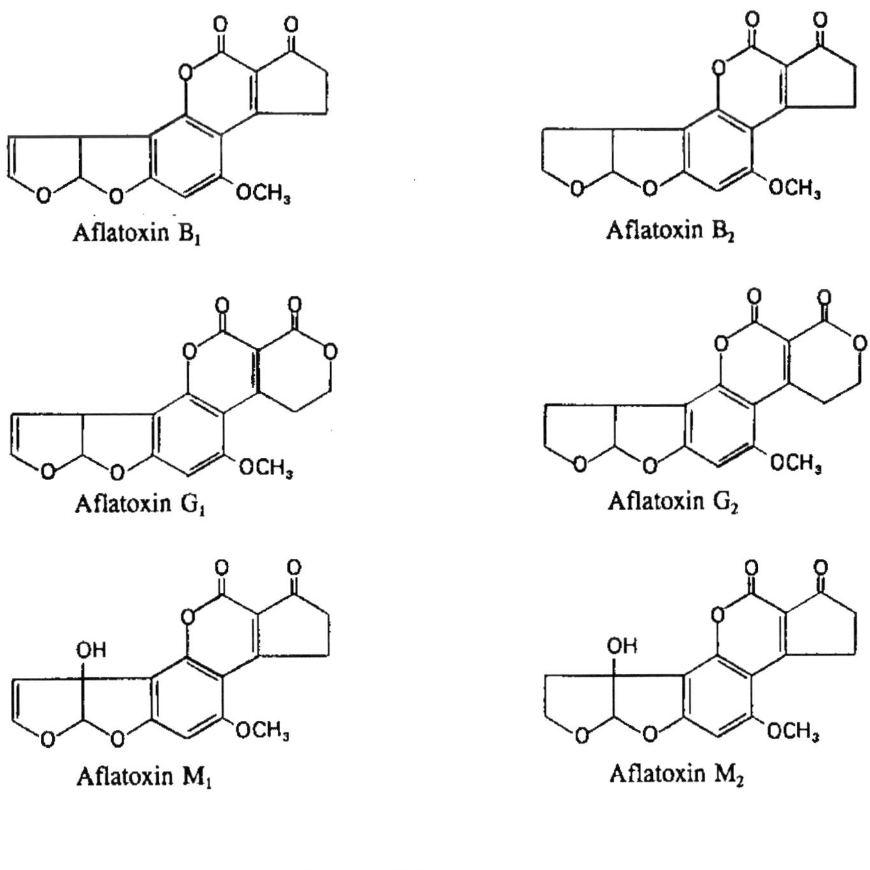

Figure 1. Chemical structure of main aflatoxins [8].

Despite the structural similarities, aflatoxins present different degrees of biological activity. AFB_1, besides being the most frequently found in cereals, presents the highest toxigenic potential, followed by G_1, B_2 and G_2 [9]. Aflatoxin B_1 is the most toxic, in acute as well as in chronic aflatoxicoses, as for aflatoxin M_1 (AFM_1), a detoxification product present in milk, as a result of the AFB_1 metabolism, is a hepatotoxic disease as potential as AFB_1 [10]. Terao and Ueno [11] have shown that the magnitude of the AFG_2, AFB_2 and AFG_1 toxicity corresponds to 10, 20 and 50% of the AFB_1, respectively.

Aflatoxins are characterized by their high toxicity. In animal health, various domestic and experimental species are sensitive to their acute toxic, mutagenic and carcinogenic effects, having the liver as the main affected organ [2]. Similarly in Public Health, aflatoxins are identified as factors involved in the hepatic cancer etiology in human, as a consequence of contaminated food ingestion [12].

Sensitivity to aflatoxins' toxic effects may significantly vary among animal species. Even among individuals within the same kind, the input-output relation may vary according to race, sex, age and diet composition, among other factors [9].

In many species, males can be more susceptible than females, although, in general, youngsters can be considerably more sensitive than adults [12]. Moreover, aflatoxins' toxic effects depend on the dose and exposure time, which determine acute or chronic poisoning. Acute toxic syndrome occurs in the ingestion of feed containing high concentration of aflatoxin, and the effects can be seen in a short time. This happens mainly due to fast deterioration of the animal's general state, loss of appetite, acute hepatitis, jaundice, hemorrhages and death [2].

In chronic aflatoxicoses, the most evident clinical sign is unthriftiness in young animals [6]. It occurs through ingestion of feed containing low levels of aflatoxin for a long term, being the exposure to the contaminant continuous or intermittent. Such pathology presents difficult diagnoses, although it represents the main poisoning form in natural conditions, which leads to significant economic losses in breeding [13].

Aflatoxins also have effects on the immune system. Among the effects of immunosuppression, demonstrated in poultry and other animals experimentation, it can be remarked aplasia of the thymus and bursa of Fabricius, reduction of the number and activity of T cells, decrease of antibody response, suppression of phagocytic activity and reduction of humoral components such as complement (C4), interferon and immunoglobulins IgG and IgA [13,14]. Such changes contribute to the occurrence of concomitant infections, especially for viral and bacterial agents associated with the exposure of animals to feed contaminated with aflatoxin

Fumonisins

Fumonisins comprise the latest group of mycotoxins discovered. Since its isolation in 1988, it has been associated with animal diseases previously known as equine leukoencephalomalacia and porcine pulmonary edema [6].

Fumonisins are produced by fungi belonging to the genus *Fusarium*. The main producer is the *Fusarium verticillioides* (=*moniliforme*), but other species of *Fusarium* are also producers, as *F. proliferatum*, *F. nygamai*, *F. anthophilum*, *F. dlamini* and *F. napiforme* [15-17]. However, fungi of the genus *Alternaria* spp. can also produce fumonisins [18].

Currently, 28 molecular structures designated by the term fumonisin are known [19, 20], but the predominant toxin produced by strains of *F. verticillioides* is fumonisin B_1 (FB_1) [21]. Only FB_1, besides fumonisins B_2 (FB_2) and B_3 (FB_3) were detected when the production of fumonisin occurred in natural conditions [22].

Studies made with rodents' hepatocytes have shown that fumonisins block the formation of sphingolipids. This evidence supports the hypothesis that the interruption of the formation of sphingolipids, resulting in accumulation of sphinganine and sphingosine, is the mechanism by which it is manifested the effects of acute toxicity and carcinogenicity of fumonisins [20]. The sphingolipids are important for maintaining cell membrane integrity and regulation of cell surface receptors, ion pumps and other systems vital to the functioning and survival of the cell [6].

Fumonisins are extremely toxic for horses and piglets, but most species of poultry shows high resistance against these toxins [6]. In a study with broilers fed with 300 mg/kg of FB_1 during the first 2 weeks of life, Brown et al. [23] it was reported diarrhea, 19% reduction in weight gain, 30% increase in liver weight as well as histological changes, such as multifocal hepatic necrosis, biliary hyperplasia, necrosis of muscle and rickets.

Immunosuppression is also seen as a major toxic effect of FB_1. Li et al. [24] found no reduction in weight gain or increased feed conversion of broilers fed with 200 mg/kg of FB_1, but reported a reduction of humoral immunity and the suppression of lymphocytes.

Zearalenone

Zearalenone is an estrogenic substance derived from resorcylate acid, produced by *Fusarium* species: *F.roseum (F.graminearum)*, *F.culmorum, F. equisetum* among others. The effects of zearalenone in animals have been known since 1974, when they were described by Mirocha and Christensen [25]. However, cases of estrogenic syndrome in swine related to this toxin were characterized by the presence of edematous vulva. Subsequently, this syndrome was detected in several countries in Europe, North America and Australia.

The most characteristic symptoms of poisoning by zearalenone are inflammation of the womb, breast and vulva in female puberty, testicular atrophy and inflammation of the breasts in young males, and, in adult animals, it causes infertility. Therefore, the effects of zearalenone are predominantly estrogenic, involving, mainly, the urogenital system. Piglets are the most commonly affected animals, however cattle, poultry, sheep and laboratory animals (rodents) are also affected [26]. In chickens and turkeys, zearalenone causes loss of body weight (bw), and production of eggs with thin shells. As for piglets, there is no weight loss unless combined with other toxins produced by *Fusarium*, like for example, trichothecenes. Among food more susceptible to contamination by zearalenone, corn can be mentioned for being an excellent substrate for fungal growth. However, other cereals can also be contaminated by the fungus, such as wheat, barley, oats, rye, sorghum and soybean [27].

Concentrations of 1 ppm or more of zearalenone in the diet can produce hyperestrogenism in piglets [28]. However, the presence of zearalenone in milk, eggs and other animal products is negligible.

Trichothecenes (DON, T-2 Toxin)

Trichothecenes are mycotoxins produced by different genera of fungi such as *Fusarium, Stachybotrys, Myrothecium, Trichotecium* among others, as these fungi are also known as pathogens of grain producing plants [29]. *Fusarium graminearum* and *F. culmorum* are the most important [27]. Currently, almost 2 hundreds compounds called trichothecenes are known, but the most studied include deoxynivalenol (DON), T-2 toxin and diacetoxisirpenol (DAS).

Studies have shown that the occurrence of DON is observed mainly in cereals such as wheat, corn, barley and oat, in concentrations up to 90 mg/kg [30]. In general, trichothecenes present in feed are easily and rapidly absorbed by the gastrointestinal tract of the exposed animal. In the case of DON, the availability of 50-60% suggests an efficient absorption. Its distribution in the body, due to polarity, is very homogeneous, neither being deposited in adipose tissue nor in other tissues and more specific organs [31]. The toxicity of trichothecenes on animals includes weight loss, feed refusal, vomiting, diarrhea, hemorrhage, and decrease in egg production, abortion and death [32].

DON is common in cereal grains in North America and Europe, where they have caused huge problems to animal health [32]. This toxin may be lethal if ingested in large quantities, and the ingestion of low-to-moderate levels may cause decreased performance and changes in the immune function [32]. Monogastric animals, particularly piglets, are highly sensitive to DON, while chickens and turkeys, followed by ruminants, seem to have high tolerance [33]. The birds are generally less sensitive to DON, since levels as high as 8 ppm in the feed has not been associated with low productivity [32].

On the recommendation of the EC (August 17, 2006), it appears that "data on the presence of T-2 and HT-2 toxins in products intended for animal feed are, so far, very limited. It is remarked, however, that there is urgency in developing and validating a sensitive method of analysis. It also warns that the possible presence of T-2 and HT-2 toxins in products intended for animal consumption is a cause for concern".

Ochratoxin A

Ochratoxins were first isolated from the species *Aspergillus ochraceus* (previously *A. alutaceus*), although they can also be produced by several species of *Aspergillus* and *Penicillium* [6].

The ochratoxins group consists of 7 components, but only ochratoxin A (OTA) has been found as a natural contaminant of grains, with worldwide distribution. The target organ of toxic action of the OTA is the kidney, interfering in the synthesis of macromolecules of the cells of the renal parenchyma, including DNA, RNA and proteins. Additionally, it affects the renal metabolism of carbohydrates, damaging the epithelium of proximal renal tubules, which decreases the absorption of electrolytes and increases the excretion of water through osmotic diuresis [6].

OTA causes various diverse effects in most domestic animals. In piglets, it acts as the causative agent of porcine mycotoxic nephropathy. Similarly, in humans, it is recognized as the etiologic agent of Balkan endemic nephropathy. For most poultry, OTA is considered the most potent mycotoxin; as they are highly lethal to animals like, for example, broilers, which

LD_{50} is around 2.1 mg/kg body weight [34]. For comparison, the LD_{50} of AFB_1 for these birds is 6.8 mg/kg body weight [35].

OTA poisoning may provide additional effects on the animals, which can cause problems, also in later stages of production. For example, Warren and Hamilton [36] observed that chickens having feed containing 2.0 mg/kg of OTA presented a significant reduction in the strength required to rupture the large intestine, besides the increase of the intestine relative weight. This fact becomes relevant when considering the possible rupture of the intestine during the poultry processing, causing contamination of the carcass by fecal material and, consequently, their conviction in slaughterhouses.

The EC issued a recommendation on August 17, 2006 on the presence of deoxynivalenol (DON), zearalenona (ZEA), ochratoxin A (OTA), T-2 and HT-2 toxins and fumonisins in products intended for animal nutrition based on scientific evidences that concluded: DON, OTA, ZEA and FBs have toxic effects in several animal species. DON, ZEA and FB_1 and FB_2 transfer, in a limited manner, from fed to the meat, eggs and milk, and food of animal origin contributes only superficially to the total exposure of humans to these toxins. OTA can transfer from feed to food of animal origin, but the exposure observance indicates that such food contribute little to the human exposure to ochratoxin A, via ingestion.

INTERACTIONS BETWEEN MYCOTOXINS

Human and animals are exposed to a variety of toxic agents that can interact in the body in different ways, altering the absorption, protein binding, biotransformation and excretion of each other. There are many possible effects from the combination of mycotoxins, causing various kinds of clinical signs. The probability of finding more than one mycotoxin in the same food is even greater as the environmental conditions are favorable to fungal development, which, as some species of the genus *Fusarium*, among others, are able to simultaneously produce more than one toxic substance. In this sense, grains of the same origin can contain more than one mycotoxin, as well as the incorporation of grains from different sources can result in food containing different mycotoxins [37].

The body responses to the combination of toxic substances may be increased or decreased as a result of the biological effects at the site of action. Several terms are used to define this interaction. An additive effect occurs when the combined action of two chemicals is equal to the sum of the effects of each agent administered alone (e.g. 2 + 4 = 6). Less than additive effects occur when the combined effect of two mycotoxins are less than the sum of the effect of each toxin given alone (e.g. 2 + 4 = 5). This is the most noticeable effect when two chemicals are administered simultaneously. The synergistic effect occurs when the combined effect is greater than the sum of the effects of each agent administered alone (e.g. 2 + 2 = 10). The term 'potentiation' is used when a product does not act in a certain system. However, when provided with another substance, it becomes more toxic (e.g. 0 + 2 = 10). As for the antagonism, it occurs when two xenobiotics, administered together, interfere in each other's action (e.g. 4 + 0 = 1) [32].

The occurrence of mycotoxicoses, due to concomitant ingestion of two or more mycotoxins manifests itself most often in livestocks, in chronic form, through ingestion of contaminated food for a long term.

Many studies have addressed the individual effect of toxins, but the interaction between the fungal metabolites in food and feed has received little attention. Most of them were conducted with birds and more rarely with piglets, sheep, rats, mice. The results have shown additive effect or less than additive in 78% of the studies; synergistic effect in 19% and antagonist effect in 3% of the experiments. The aflatoxin was one out of two mycotoxins involved in five out of six cases of synergism [32].

Huff and Doerr [34] evaluated the effects of aflatoxin and ochratoxin in poultry. After three weeks of administration the authors reported splenic, renal pro-ventricular and pancreatic hyperplasia, showing nephropathy as the primary effect of the interaction. Liver hyperplasia was only observed in chickens that received only aflatoxins. It also presented shrinkage of the liver.

In 1996, Bacon et al [38] suggested that studies on mycotoxicology with *Fusarium* toxins, which attributed the toxicity to a single compound in animals fed with naturally contaminated feed, should be re-examined in light of the interactions potential with fusaric acid.

The synergism between deoxynivalenol (DON) and fusaric acid was shown by Smith et al [39]. The authors observed an increase in dose-dependent toxicity of fusaric acid, in the administration of DON at a dose of 2.4 mg/g of food. Thus the authors remarked the discrepancy of results achieved in previous studies regarding the toxicity of DON in animals fed with naturally contaminated feed when compared to equivalent doses of purified toxin [33, 40, 41].

In 1998, studies by Smith and Seddon [42], using piglets intoxicated with a combination of deoxinivalenos (DON) and T-2 toxin showed to cause anorexia, vomiting, immune suppression, lethargy and ataxia.

One of the most studied synergisms has been the interaction between ochratoxin A and citrinin, recognized as nephrotoxic toxins. In Beagle dogs, the severity of clinical signs was demonstrated as well as mortality of the dogs that ingested both toxins. Severe renal lesions were observed accompanied by glycosuria and proteinuria [43].

Regarding the association between ochratoxin A and T-2 toxin in pregnant rats, there was fetal death. However, when administered alone at doses of 2 to 4 mg/kg bw of ochratoxin A and 0.5 mg/kg bw of T-2 toxin, pregnancy occurred naturally. It was also observed deformations on the heads of animal groups that received ochratoxin A, and tail and limbs deformation in those animals that were administrated only with T-2 toxin [44].

Potentiated toxic effects have already been demonstrated in FB_1 and T-2 toxin [45], moniliformin and aflatoxin [46] and deoxynivalenol and moniliformin [47]. Despite the frequency of *Aspergillus flavus* and *Fusarium verticillioides* and co-occurrence of aflatoxin and fumonisin in the same food, mainly in maize and derivatives, the biological effects in animals resulting from the combination of both toxins are also poorly studied and restricted mainly to piglets, birds, rodents, fish and rabbits.

Kubena et al. [45] conducted a study of the FB_1 effects in combination with aflatoxins, in turkeys, providing the birds with feed containing 200 mg of FB_1/kg and 750 mg of aflatoxin/kg (79% AFB_1, 16% AFG_1, 4% AFB_2, 1% AFG_2). The birds having feed containing FB_1 showed a 10% reduction in weight gain in relation to the control group. The birds treated with aflatoxin were reduced by 39%, and the birds treated with both mycotoxins were reduced by 47%. Furthermore, it was observed a significant decrease in relative liver weight in birds fed with both toxins. The authors concluded that the effects of FB_1 and aflatoxin, when combined, can be more severe than when they are isolatedly present. Previously, Weibking et al. [48] obtained similar results in a study conducted with young turkeys.

The effects of prolonged oral administration (21 days) of fumonisin B_1 and aflatoxin B_1 were evaluated on male laboratory rats. The results showed that treated animals presented differences in body weight and absolute/relative weights of liver and kidney as well as altered hepatic functions and cholesterol blood levels. Rats fed with the greatest doses of AFB_1 and FB_1 gained less weight at the end of the experiment period; their blood concentrations of liver and enzymes aspartate aminotranferase (AST) and alkaline phosphatase (AP) were above control levels. Blood cholesterol increased in the groups treated with the highest dose of FB_1 or FB_1 associated with AFB_1. Histopathology revealed the occurrence of apoptosis in the liver of rats exposed to FB_1. The association of aflatoxin B_1 with fumonisin B_1 at a higher dose probably potentiated the effects of the higher dose of fumonisin B_1 acting alone [49].

Casado et al. [50] conducted a study on the association of the effects of AFB_1 and FB_1 in the diet of mice for 90 days. Three experimental groups were formed, as the first group received 10 mg/kg of FB_1 in their diet, the second group 10 mg/g of AFB_1 and the third group 10 mg/kg of FB_1 + 10 mg/g of AFB_1. It was observed that on the thirtieth day of the trial, no change was detected. On the sixtieth day, the mice in the control group had no liver injury, while 50% of the sacrificed animals treated with AFB_1 and FB_1 had points of inflammation and necrosis in the liver and small intestine. About 70% of mice treated with both toxins presented hyperplasia of Kupffer cells in the liver. In the ninetieth day of the trial, these same changes were found, but in a percentage of 80% in animals treated with AFB_1, 90% in those treated with FB_1 and 100% of mice treated with both toxins. There were no significant changes in body weight or weight gain of these animals.

Carlson et al. [51] conducted an experiment with rainbow trouts, studying the effects of FB_1 when co-administered with AFB_1. At the end of the study, it was revealed that the FB_1 administered singularly (0, 3.2, 23 or 104 mg / kg) for 34 weeks did not cause liver cancer in the trouts, but when administered with AFB_1 (23 or 104 mg / FB_1 kg + 100 mg/kg of AFB_1) for 42 weeks, there was a significant increase in the incidence of liver cancer in relation to control group and other groups.

Dilkin et al. [52] studied the toxicological effects of chronic oral exposure (28 days) to aflatoxin B_1 and fumonisin B_1 in weaned piglets. Six experimental groups were fed with rations containing: (A) 0 mg of FB_1 and 0 μg of AFB_1; (B) 10 mg of FB_1/kg of feed; (C) 30 mg of FB_1/kg of feed; (D) 50 μg of AFB_1/kg of feed; (E) 10mg of FB_1 plus 50 μg of AFB_1/kg feed; (F) 30 mg of FB_1 plus 50 μg of AFB_1/kg of feed. The results showed that all piglets from groups C and F presented typical signs of pulmonary edema, with reduced feed consumption and body weight gain as well as pathological alteration. FB_1 was detected in feces and urine at 24 h of poisoning and in liver after 28 days of intoxication. Increases were detected regarding the following hematological and biochemical parameters in animals from treatments C and F: erythrocyte number; hematocrit; total bilirrubin; total protein; activity of serum alkaline phosphatase, aspartate aminotransferase, and alanine aminotransferase. Cholesterol levels significantly augmented only in animals from groups C and F. Whereas, albumin concentrations increased in groups C, F, B and E. The average organ/body weight ratio of piglets (heart, liver and lung) were significantly greater in groups C and F. Chronic intoxication of piglets with AFB_1 and FB_1 leads to important losses of productivity.

The effects of prolonged oral administration (21 days) of fumonisin B_1 (FB_1) and aflatoxin B_1 (AFB_1) were studied in male New Zealand rabbits by clinical, pathological, biochemical and sphingolipid analyses. Differences were observed between intoxicated rabbits and controls with respect to absolute and relative liver and kidney weight, hepatic function, serum urea and creatinine levels and Sa/So ratio. The most frequent hepatic and renal injuries were vacuolar degeneration of the liver and kidney as shown by the histopathological and serum biochemical results. The effects of FB_1 administered alone were dose dependent, with the observation of a more intense hepatic effect compared to the renal effect at the highest dose used. On the other hand, combined administration of AFB_1 and FB_1 resulted in synergistic toxic effects both in the liver and in the kidney, but hepatic injuries were more marked [53].

Conclusion

The probability of finding more than one mycotoxin in the same food is even greater as the environmental conditions are favorable to fungal development. The occurrence of mycotoxicoses, due to concomitant ingestion of two or more mycotoxins manifests itself most often in livestocks, in chronic form, through ingestion of contaminated food for a long term. Toxicity of some individual mycotoxins can be enhanced in a synergistic, additive or antagonistic manner when they occur as co-contaminants and are consumed by different animal models.

References

[1] Diaz, GJ; Squires, EJ; Julian, RJ; Boermans, HJ. Individual and combined effects of T-2 toxin and DAS in laying hens, *British Poultry Science,* 1994, 35, 473–485.

[2] Osweiler, GD. Mycotoxins and livestock: What role do fungal toxins play in illness and production losses? *Veterinary Medicine*, 1990, 85, 89-94.

[3] Clevstrom G. Studies on the Fungal Flora of Plants and Feed and the Influence of Formic Acid on Growth and Aflatoxin Production in *Aspergillus flavus*. PhD Thesis. Swedish University of Agricultural Sciences, Department of Microbiology. 1986, 15-19.

[4] Frisvad, JC; Samson, RA. Filamentous fungi and feeds: ecology, spoilage and mycotoxins productions. In: Dilip K; Arora KG.; Mukerji, E; Marth, H (Eds.) *Handbook of Applied Mycology, Food and Feeds*. Marcel Dekker Inc. New York, 1991, 31-68.

[5] Burgess, L; Nelson, PE; Tousson, TA. Laboratory manual of *Fusarium* Studies, Sydney: University of Sydney: 1981.

[6] Leeson, S; Diaz, GJ; Summers, JD. *Poultry metabolic disorders and mycotoxins*. Guelph: University Books: 1985.

[7] Hussein, HS; Brasel, JM. Toxicity, metabolism, and impact of mycotoxins on humans and animals. *Toxicology,* 2001, 167, 101-134.

[8] Smith, JE; Moss, MO. Mycotoxins: formation, analysis and significance. Chichester: J. Wiley: 1985.

[9] Coulombe, RA. Aflatoxins. In: Sharma, RP; Salunke, DK. *Mycotoxins and phytoalexins*. Boca Raton: CRC Press: 1991, 103-143.
[10] Carnaghan, RBA.; Hartley, RD; Okelly, J. Toxicity and fluorescence properties of the aflatoxins. *Nature,*, 1963, 200, 911-1101.
[11] Terao, K; Ueno, Y. Morphological and functional damage to cells and tissues. In: Uraguchi, K; Yamazaki, M. *Toxicology, Biochemistry and Pathology of Mycotoxins*. Kodansha Press; Tokyo; 1978; 189-238.
[12] McLean, M; Dutton, MF. Cellular interactions and metabolism of aflatoxin: an update. *Pharmacology and Therapeutics,* 1995, 65, 163-192.
[13] Pier, AC. Major biological consequences of aflatoxicosis in animal production. *Journal of Animal Science*, 1992, 70, 3964-3967.
[14] Petska, JJ; Bondy, GS. Alteration of immune function following mycotoxin exposure. *Canadian Journal of Physiology and Pharmacology*, 1990, 68, 1009-1016.
[15] Shephard, GS; Thie, PG; Sydenham, EW. Initial studies on the toxicokinetics of fumonisin B_1 in rats. *Food and Chemical Toxicology,* 1992, 30, 277-279.
[16] Ross, PF; Nelson, PE; Richard, ID; Osweiller, GD; Rice, LG; Plattner, RD; Wilson, TM. Production of fumonisins by *Fusarium moniliforme* and *F. proliferatum* isolates associated with equine leukoencephalomalacia and pulmonary edema syndrome in swine. *Applied and Environmental Microbiology,* 1990, 56, 3225-3226.
[17] Nelson, PE. Taxonomy and biology of *Fusarium moniliforme*. *Mycopathologia*, 1992, 117, 29-36.
[18] Chen, J; Mirocha, CJ; Xie, W; Hogge, L; Olson, D. Production of the mycotoxin fumonisin B_1 by *Alternaria alternata f. sp. lycopersici. Applied and Environmental Microbiology*, 1992, 58, 3928-3931.
[19] Musser, SM; Plattner, RD. Fumonisin composition in cultures of *Fusarim moniliforme, Fusarium proliferatum* and *Fusarium nygami. Journal of Agricultural and Food Chemistry*, 1997, 45, 1169-1173.
[20] Ah-Seo, J; Won Lee, Y. Natural occurrence of the C series of fumonisins in moldy corn. *Applied and Environmental Microbiology,* 1999, 65, 1331-1334.
[21] Norred, WP. Fumonisins - mycotoxins produced by *Fusarium moniliforme*. *Journal of Toxicology and Environmental Health,* 1993, 38, 309-328.
[22] Henry, MH; Wyatt, RD. A review of fumonisin production by *Fusarium moniliforme* and fumonisin toxicoses in animals. *Journal of Applied Poultry Research*, 1993, 2, 188-192.
[23] Brown, TP; Rottinghaus, GE; Williams, ME. Fumonisins mycotoxicosis in broilers: performance and pathology. *Avian Diseases*, 1992, 36, 450-454.
[24] Li, YC; Ledoux, DR; Bermudez, AJ; Fritsche, KL; Rottinghaus, GE. Effects of fumonisin B_1 on selected immune responses in broiler chicks. *Poultry Science,* 1999, 78, 1275-1282.
[25] Mirocha, CJ; Christensen, CM. Oestrogenic mycotoxins synthesized by *Fusarium*, In: Mycotoxins. Amsterdam, J.F.H. Purchase, Ed. Elsevier: 1974, 139-149.
[26] Hagler, WM Jr; Towers, NR; Mirocha, CJ; Eppley, RM; Bryden, WL. Zearalenone: mycotoxin or mycoestrogen? In: Summerell, BA; Leslie, JF; Backhouse, D; Bryden WL; Burgess LW (Eds.) *Fusarium.* Paul E. Nelson Symposium. American Phytopathological Society Press, St. Paul, Minn: 2001, 321-331.

[27] Pittet, A. Natural occurrence of mycotoxins in foods and feeds -an updated review. In: Mycotox 98. Mycotoxin in the Food Chain. Processing and Toxicological aspects. vol. 149. Bars, JL; Galtier, P (Eds.). *Revue de Medecine Veterinaire*, Toulouse, France: 1998, 479-492.

[28] Kurtz, HJ; Mirocha, CJ. Zearalenone (F-2) induced estrogenic syndrome in swine. In: Wyllie, T.D.; & Morehouse, L.G. (Eds.) *Mycotoxic Fungi, Mycotoxins, Mycotoxicoses: an Encyclopedia Handbook.*, Dekker, New York, 1978, 256-268.

[29] Miller, JD. Fungi and mycotoxins in grain: implications for stored product research. *Journal of Stored Products Research,* 1995, 31 1–6

[30] WHO (World Health Organization). Mycotoxins. Geneva, WHO, 1990.

[31] Galtier, P; Paulin, F; Eeckhoutte, C; Larrieu, G. Comparative effects of T-2 toxin and diacetoxyscirpenol on drug metabolizing enzymes in rat tissues. *Food and Chemical Toxicology,* 1989, 4, 215-20.

[32] CAST (Council for Agricultural Science and Technology). Mycotoxins: Risks in Plant, Animal, and Human Systems. Task Force Report no.139, 199p. Ames, Iowa, USA, 2003.

[33] Prelusky, DB; Savard, ME; Trenholm, HL. Pilot study on the plasma pharmacokinetic fate of fumonisin B_1 in cows following a single dose by oral gavage or intravenous administration. *Natural Toxins*, 1995, 3, 389-394.

[34] Huff, WE; Doerr, JA. Synergism between aflatoxin and ochratoxin A in broiler chickens. *Poultry Science*, 1981, 60, 550-551.

[35] Smith, JW; Hamilton, PB. Aflatoxicosis in the broiler chickens. *Poultry Science*, 1970, 49, 207-215.

[36] Warren, MF; Hamilton, PB. Intestinal fragility during ochratoxicosis and aflatoxicosis in broiler chickens. *Applied and Environmental Microbiology*, 1980, 40, 641-645.

[37] Trenholm, HL; Prelusky, DB; Young JC; Miller JD. A practical guide to the prevention of *Fusarium* mycotoxins in grain and animal feedstuffs. *Archives of Environmental Contamination and Toxicology,* 1989, 18, 443–451.

[38] Bacon, CW; Nelson, PG. Fumonisin production in corn by toxigenic strains of *Fusarium moniliforme* and *Fusarium proliferatum. Journal of Food Protection*, 1994, 57, 514-521.

[39] Smith, TK.; McMillan, EG; Castilho, JB. Effects of feeding blends of *Fusarium* mycotoxin-contaminated grains containing grains containing DON and fusaric acid on growth and feed consumption of immature swine. *Journal of Animal. Science*, 1997, 75, 2184-2191.

[40] Rotter, BA; Thompson, BK; Trenholm, HL. Influence of low-level exposure to *Fusarium* mycotoxins on selected immunological and hematological parameters in young swine. *Fundamental and Applied Toxicology,* 1994, 23, 117-124.

[41] Seddon, IR; Smith, TK; De Lange, CMF. *Fusarium* mycotoxicoses in weaner and starter pigs. In: *Ontario Swine Research Review,* 1997.

[42] Smith, TK; Seddon, IR. Synergism demonstrated between Fusarium mycotoxins. 1998. *Feedstuffs, June 22nd,* 12-17.

[43] Kitchen, DN; Carlton, WW; Tuite, J. Ochratoxin A and Citrinin induced nefrosis in Beagle dogs. Clinical and clinicopathological features. *Veterinary Pathology*, 1997, 14, 154 – 172.

[44] Hood, RD; Kucsuk, M; Szcech, GM. Effects in mice of simultaneous prenatal exposure to ochratoxin A and T-2 toxin. *Teratology,* 1978, 17, 25-29.

[45] Kubena, FF; Edrington, TS; Kamps-Holtzzapple, C; Harvey, RB; Elissalde, MH; Rottinghaus, GE. Effects of feeding fumonisin B_1 present in *Fusarium moniliforme* culture material and aflatoxin single and combination to turkey poultry. *Poultry Science,* 1995, 74, 1295-1303.

[46] Kubena, LF; Harvey, RB; Buckley, SA; Bailey, RH; Rottinghaus, GE. Effects of long-term feeding of diets containing moniliformin supplied by *Fusarium fujikuroi* culture material, and fumonisin, supplied by *Fusarium moniliforme* culture material, to laying hens. *Poultry Science,* 1999, 78, 1499-1505.

[47] Harvey, RB; Kubena, LF; Rottinghaus, GE; Turk, JR; Casper, HH; Buckley, SA. Moniliformin from *Fusarium fujikuroi* culture material and deoxynivalenol from contaminated wheat incorporated into diets of broiler chicks. *Avian Diseases,* 1997, 41, 957-963.

[48] Weibking, TS; Ledoux, DR; Bermudez, AJ; Turk, JR; Rottinghaus, GE. Effects of feeding *Fusarium moniliforme* culture material, containing known levels of fumonisin B_1, on the young broiler chick. *Poultry Science*, 1993, 72, 456-466.

[49] Pozzi, CR; Correa, B; Xavier, JG; Direito, GM; Orsi, RB; Maraeazzo, SV. Effects of prolonged oral administration of fumonisin B_1 and aflatoxin B_1 in rats. *Mycopathologia*, 2000, 151, 21-27.

[50] Casado, JM; Theumer, M; Masih, DT; Chulze, S; Rubinstein, HR. Experimental subchronic mycotoxicoses in mice: individual and combined effects of dietary exposure to fumonisin and aflatoxin B_1, *Food and Chemical Toxicology*, 2001, 39, 579–586.

[51] Carlson, DB; Williams, DE; Spitsbergen, J M; Ross, PF; Bacon, CW; Merideth, FI; Riley, RT. Fumonisin B_1 promotes aflatoxin B_1 and N-methyl-N′-nitronitrosoguanidine-initiated liver tumors in rainbow trout. *Toxicology and Applied Pharmacology,* 2001, 172, 29-36.

[52] Dilkin, P; Zorzete, P; Mallmann, CA; Gomes, JDF; Utiyama, CE; Oetting, LL; Correa, B. Toxicological effects of chronic low doses of aflatoxin B_1 and fumonisin B_1-containing *Fusarium moniliforme* culture material in weaned piglets. *Food and Chemical Toxicology,* 2003, 41, 1345-1353.

[53] Orsi, RB; Oliveira, CAF; Dilkin, P; Xavier, JG; Direito, GM; Correa, B. Effects of oral administration of aflatoxin B_1 and fumonisin B_1 in rabbits (*Orictolagus cuniculus*). *Chemico-Biological Interactions*, 2007, 170, 201-208.

In: Mycotoxicoses in Animals Economically Important ISBN: 978-1-61668-195-1
Editors: Edlyane Gonçalez, et. al. pp. 131-142

Chapter 6

NUTRITIONAL EFFECTS OF MYCOTOXINS: IS THERE ANY PROPERTY MANAGEMENT?

Simone Aquino[1*], Edlayne Gonçalez[2] and Joana D'arc Felicio[2]
[1] Codeagro, Av Miguel Estéfano, 3.900, São Paulo, SP, Brazil, Cep 04301-903
[2] Centro de Pesquisa e Desenvolvimento de Sanidade Animal, Instituto Biológico, Av. Cons. Rodrigues Alves, 1252, São Paulo, SP, Brazil, CEP 04014-002

ABSTRACT

Moulds and associated mycotoxins are important factors adversely affecting foods produced using contaminated plant products or contaminated feeds that impair farm operations as well as feed production in various ways. Mycotoxins of agroeconomic significance include aflatoxins, ochratoxins, trichothecenes, zearalenone, fumonisins, patulin, moniliformin, cyclopiazonic acid and ergot alkaloids. Mycotoxins are invisible and cannot be detected by smell or taste, but can reduce performance in animal production significantly. The occurrence of mould and mycotoxins in food and animal feed is a major problem globally. The impact of mycotoxins on the health of farm animals has involved account for millions of dollars annually in losses world-wide in animal health including loss of animal life and veterinary care costs. This chapter demonstrated the influence of nutrients levels such as protein, vitamins and minerals on animal recovery and how the use of absorbents interfere on the balance of nutrients necessary to growth or animal performance.

INTRODUCTION

There is general agreement that an increased microbial activity in feeds may affect both their chemical composition and nutritional value. Chemical changes in feeds caused by moulds, such as a lower fat content, increased fat acidity, lowered protein content, changes in

* Corresponding author: S. Aquino
Telephone: +55-11-50871753; Fax: +55-11-50871754
E-mail address: siaq06@hotmail.com

the content of amino acids and free fatty acids, and changes in the content of non-reducing sugars, have been described [1,2].

Many mycotoxin-producing fungal species can cause plant injury under field conditions and mycotoxins themselves play a role in plant diseases in addition to their role in animal diseases. Some mycotoxins are highly toxic to plants and the increase production of a mycotoxin increases the ability of a fungus to cause plant damage. Besides, assessment of the nutritional impact of fungi degradation on the health of farm animals requires a full understanding. How do fungi degrade and obtain nutrients from substrate or plant? Some strategies have been used to investigate *Aspergillus* virulence using mutants. The virulence of fungi can be caused by the production of fungal proteins (enzymes) that promote mycelial growth, based on the biochemical function of the enzyme (e.g. protease, oxidase, phospholipase), which, potentially, could help the invading mycelium to grow [3,4].

According to Isaac [3] the degradation occurs as the result of the production of appropriate extracellular enzymes and a large number of fungal species produces such enzymes. The degradative process is complex and occurs as the result of several biochemical steps. Fungi require sources of organic carbon for growth and development. Most organic carbon occurs as carbohydrates bound into the major components (in complex forms) of plant tissues. In particular, the wide range of enzymes produced by *Aspergillus* for the degradation of plant cell wall polysaccharides is of major importance to the food and feed industry. Recently, several *Aspergillus* spp. have received increased interest as hosts for heterologous protein production [4].

Fungi are efficient cellulose degraders and sufficient water is available to the penetrative nature of hyphal growth ensures that extracellular enzymes quickly reach into plant tissues. Cellulose is a major component of plant and cellulose fibrils are degraded by a series of enzyme reactions and in this way glucose is liberated from cellulose microfibrils and can be absorbed into fungal hyphae [3]. This effect represents a competition with mould for energy sources available to the animals and plasma glucose levels could be already reduced, as observed in calves receiving aflatoxin, because liver is the primary organ affected by aflatoxin [6].

Dänicke et al. [7] studied diets to fattening pigs containing the DON-contaminated wheat with supplementation of a detoxifying agent. The digestibility of components was significantly higher in diets containing the DON-contaminated wheat than in the control diets. This controversial result with significantly higher feeding value of contaminated diets could have several explanations. It becomes clear that the *Fusarium*-induced alterations in cereal grains not only include an accumulation of mycotoxins but might have substantial impacts on its feeding value. *Fusarium culmorum* excretes cell wall degrading enzymes such as xylanases, pectinases and cellulases when infecting wheat ears and increased activities of amylase, protease and non-starch polysaccharide-degrading enzymes in the harvested wheat kernels when compared to the control wheat. Moreover, the *Fusarium* invasion resulted in an increased protein and amino acid concentration, a decreased starch concentration and decreased water extract viscosity. Hence, it could well be that the growth of several *Fusarium* fungi facilitates the digestion of cereal nutrients by non ruminants via partial cell wall degradation. It was suggested that fungus protein with its own high biological value could improve the protein quality of the infected cereal [8].

As the information on the effect of moulds on the nutritional value of animal feeds is limited, Blaha et al. [9] studied the effects of inoculated wheat with three storage mould strains. The results of the balance experiments showed that wheat heavily infested with *Penicillium cyclopium, Mucor* spp. and *Aspergillus flavus,* had an unpleasant odour, but the true digestibility of protein (%) and biological value of its protein in rats was not significantly affected. In despite of changes in its sensory quality and lipid components, no significant adverse effects on the utilization of protein by the rats were observed. It seems that the rats are capable of utilizing fully the protein of a mould-infested feed, even though part of it had been obviously metabolized. The energy losses caused by the metabolic activity of moulds were probably not serious enough to affect the results of the experiments. The results of feeding experiments on rats, ducklings and broiler chicks fed with rations containing 5-50% of mouldy grain heavily invaded by a mixture of naturally present fungi plus *A. flavus* NRRL 2999 did not show any injury to the animals [10]. As for the study of the effect of storage duration on changes in the nutritional value of animal feeds, there are similarities and controversial results.

Stringhini et al. [11] evaluated the use of low nutritional quality corn grain, infested with fungi on the performance of broilers in starter diets. The non infested corn was used in the control diet, and substituted by levels of 20 or 40% of grains infested by molded grains. In the necropsy done at the 29 and 59 days of broiler age, there was an increase on the frequency of leg problems and hepatic lesions in the treatments containing corn grain infested by mould in its composition. The contaminated grains used did not interfere with the broiler performance, however it was observed that the increase of the incidence of metabolic problems did not reflect in the broiler performance, but it increased the carcass condemnations. Obviously moldy feed should be avoided. Spoilage or deteriorated silage can reduce feed consumption, fiber digestibility and production [12, 13].

The known dietary factors that interact with mycotoxins include nutrients such as fat, protein, fiber, vitamins and minerals [14-16]. Dietary aflatoxin caused in broiler chickens a significant decrease in the specific activities of pancreatic enzymes. These enzymes are the primary enzymes of digestion for starches, protein, lipid and nucleic acids. Clinical signs in outbreaks can vary or be different comparing to experimental findings in animals dosed with mycotoxins. The malabsorption of various nutrients was observed in farm animals, including poor growth as an effect of mycotoxicosis in animals. In field outbreaks aflatoxicosis, the most prominent feature is the low absorption of nutrients, leading to the appearance of particles of feed in the feaces associated with *steatorrhea* or increase of lipids excretion [17].

The deterioration of the nutritional value of some moulded feeds was reported by Whiting [18], Sharby et al. [19] and the Food and Agriculture Organization [20]. It has been shown that the oxidation of lipids gives rise to various types of carbonyl derivatives reacting with free amino acids, and amine groups of basic amino acids bound in proteins, with the production of Schiff bases. The carbonyl derivatives react with free NH_2- groups of protein-bound lysine, with lysine being transformed into enzyme resistant compounds [9]. *Aspergillus fumigatus* produces several mycotoxins, including gliotoxin that are toxic to cattle and affect rumen fermentation, reducing digestibility. Besides, cattle consuming this silage demonstrated symptoms including generalized deterioration typical of protein deficiency, malnutrition, diarrhea, irritability, abnormal behavior and occasional death [21,22].

The relationship of aflatoxins and amino acids is not clear. The supplementation with 4% methionine, 1% arginine or 0.8% of lysine in diet decreased the mortality of ducks, but decreased the weight also [23]. It was observed that the addition of glutathione or its precursors amino acids in feed decreased the severity of aflatoxicosis, once the detoxification of aflatoxin occurs with the conjugation of reduced glutathione with 2,3 epoxy [24]. The supplementation with glutathione or cysteine re-established the feed consumption of broilers to a normal level [25]. On broilers diet it was observed that the negative effect of aflatoxin on growth rate was major on deficient diet (66%) than adjusted diet (100%), but no positive effect was observed in diets with 134% of methionine. It suggested that methionine had a little effect on aflatoxin [26].

The interrelationships between mycotoxins and the utilization of dietary protein were reviewed and data demonstrated that most mycotoxins impair utilization of dietary protein. Initially, some studies showed that the symptoms of aflatoxicosis were most severe in broilers, rats and swines when were fed with a protein poor diet comparing with those animals feed with a normal protein diet [27]. Inversely, the diets with high levels of protein demonstrated a protector effect against aflatoxins in broilers [28]. These findings corroborate that aflatoxins in feed increase the necessity of proteins to obtain a minimum of productivity and this is an important consideration once proteins are very expensive macro nutrients and animal growth is limitated according to proteins levels [17].

The effectiveness of protein supplements in overcoming mycotoxicoses will depend on the mycotoxin in question. Results obtained with poultry and swine have shown that supplements of dietary protein beyond normal requirements can overcome these conditions. High-protein diets, however, have been shown to promote hepatoma characteristic of chronic aflatoxicosis in rats. Aflatoxin interferes with utilization of dietary protein by inhibiting synthesis of DNA, RNA, and protein. High-protein diets promote the metabolism of aflatoxin by the hepatic microsomal drug-metabolizing enzyme system. It was observed that zearalenone increased membrane permeability and promotes uterine synthesis of DNA, RNA, and protein. Supplements of dietary protein overcame growth reduction and reduced the metabolic half-life of the toxin by promoting urinary excretion of free, unmetabolized zearalenone in the rat. The trichothecene mycotoxins can disrupt normal protein metabolism by inactivating the ribosomal cycle. Protein supplements appear to have little effect on trichothecene mycotoxicoses [29].

Verma et al. [30] evaluated the effect of dietary aflatoxin B_1 (AF) at levels of 0.5, 1 and 2 mg/kg, ochratoxin A (OA) at levels of 1, 2 and 4 mg/kg and their corresponding combinations on protein and energy utilization as well as energy partitioning in white leghorn laying hens. Protein retention and metabolizable energy (ME) value of the diets was adversely affected at all levels of AF and OA either singly or in combination, though the effect was more evident with OA and AF + OA. Minimum protein and ME value retention was recorded in hens fed the combination of toxins at their highest levels (2 mg/kg AF + 4 mg/kg OA). It is suggested that AF and OA either singly or in combination affect not only protein and energy utilization in laying hens but also energy partitioning i.e. egg and body energy deposition and maintenance energy requirement. However, the combination of toxins (AF + OA) has more severe adverse effects on all parameters than the individual toxins because of their synergistic toxicity effect.

The interaction of aflatoxin with vitamins is not clear in chickens. The supplementation in four times more than levels recommended by National Research Council (NRC) showed no protector effect against aflatoxins on broilers growth [31]. However, experimental work has revealed a significant interaction between aflatoxin and colecalciferol (vitamin D) with respect to bone calcification in chicks, this increasing their vitamin D requirements. Paleness of the legs and mucosas is also often observed in broilers and laying hens. The aflatoxicosis is known as "pale bird syndrome" as a consequence of minor absorption, decreasing of transport and tissue deposition of carotenoids in diet [32].

Deficient levels of riboflavin or colecalciferol decreased the development of broilers, even with low concentration of aflatoxins. Diets deficient in riboflavin and colecalciferol may significantly increase the susceptibility of broilers to aflatoxins, resulting in a poor growth performance. The adverse effects of aflatoxins are more intense during the initial periods of growth, i.e. when the animals are fed contaminated food during the first 21 days of life [33].

The opposite effect was observed when the vitamin B_6 deficient showed a better performance in intoxicated broilers. The cause is explained by deficient B_6 levels that rise of fatty acids oxidation in excess on the liver, deposited by aflatoxin action [32]. Chickens with aflatoxicosis symptoms should received riboflavin and vitamin D on feed and decrease on diet the levels of B_6 vitamin. Bryden et al. [34] observed that vitamin A (retinol) increases mortality rate in chicks intoxicated with aflatoxins when fed for more than 2 weeks. Aflatoxin increased relative liver weight, liver lipid concentration and reduced growth; however, these effects were less pronounced with avitaminotic A chickens.

The molecular mechanism of action of vitamin E on mammalian cells was studied. Vitamin E dietary intake was assessed for its effects on the initiation phase of carcinogenesis. Rats were fed *ad libitum* a diet containing 0, 0.05, 0.5 or 5 IU of alpha-tocopherol for 8 weeks. Modulation of vitamin E level in post mitochondrial and microsomal fractions resulted in nutritional effects. Cytochrome P-450 content was not modified by the level of vitamin E in the diet. The microsomal P-450 activities, P-450 IIB1 and IIIA, were decreased in the deficient group to -35% and -16%, respectively, as compared with control diet (0.05 IU). Diet supplemented with 0.5 IU of vitamin E increased P-450 IIB and IIIA activities (+28% and +37%, respectively) whereas a diet highly supplemented in vitamin E (5 IU) reduced these specific P-450 activities. Lipid peroxidation, estimated by the formation of thiobarbituric acid reactive products, increased in the dietary vitamin E free diet (+20%) and strongly decreased in the supplemented group (-99%). This study establishes that in vivo, dietary vitamin E protects directly membrane against damage induced by lipid peroxidation and indirectly hepatic microsomal monooxygenase activities. However, vitamin E accumulation seems to alter membrane structure and function [35].

Ramos et al. [36] observed alterations of main minerals constituents (Ca, P, Mg, Na, K, Cu, and Zn) of blood in lambs intoxicated with 2.5 ppm of aflatoxins. Samples of serum and plasma were obtained at 0, 7, 14, and 21 days of intoxication. A decrease in plasma Zn was observed at the 14th day of intoxication. Ca, P, Mg, and K levels were lower by Day 21. Cu levels increased compared with controls. Na levels remained in the control range of lambs. During the clearance time, Cu level returned to control values by Day 4, Ca and K by Day 8, whereas P, Mg and Zn remained lower. These results indicated that during aflatoxicosis in lambs a severe alteration in the mineral metabolism or absorption was produced, mainly due to hepatic and renal lesions and the decrease in feed intake observed in this study.

The aflatoxins are considered unavoidable contaminants of foods. There is no specific treatment for mycotoxicosis, due to a variety of symptoms and recovery is unlikely in acute mycotoxicosis. In all cases, animals with suspected mycotoxicosis should be removed from the feed responsible. If unacceptably high levels of mycotoxins occur, dilution or removal of the contaminated feed is preferable; however, it is often impossible to completely replace some feeds in the ration, particularly the forage ingredients. Chemical, physical, and biological methods are reviewed along with other methods for management and detoxification of mycotoxins. Chemoprevention, chemoproctetion, or chemoprophylaxis are terms that describe chemical or nutritional dietary interventions that can alter the susceptibility of animals and the potential health hazards associated with aflatoxin exposure [37].

Ammonia and formaldehyde have been used as reactive chemicals for detoxifying aflatoxin, ochratoxin, and zearalenone in grain. Treatment with chemical preservatives may change the chemical composition of the substrate. As an example, formaldehyde reacts with protein and causes decreased digestibility of crude protein and depressed available lysine in treated barley. Ammonia also reacts with protein, and toxic imidazoles may be produced by reaction with sugar. The vitamin E content of grain is decrease after treatment with ammonia, propionic or formic acid [38]. Cleaning or ammoniation of grains can reduce mycotoxin concentrations, but there is no practical method to detoxify affected forages [12,13].

The physical process, through the use of mixed adsorbents the animal feed is used currently. The adsorbents combine with mycotoxins, therefore its absorption in the intestine of the animals, disabling its distribution for the target site. The main types of adsorbents are: hydrated sodium calcium aluminosilicate, activated carbon, cholestyramine, polymeric glucomannan and zeolites. It was possible to observe that the adsorbents are efficient against the mycotoxins. But it must be standed out that still adsorbents does not exist model, a time that is limited for mycotoxins determined and for adsorbent nutrient of the animal feed. The results demonstrated "in vitro" tests will not have the same alive effectiveness necessarily "in vivo" tests [39].

A growing experiment and a digestibility study were carried out with fattening pigs [7]. Diets containing the deoxynivalenol (DON)-contaminated wheat were compared to control diets and additionally, these diets were fed either without or with supplementation of a detoxifying agent (Mycofix®Plus -MP). Pig diets with DON concentrations of approximately 2.7 mg/kg (starter) and 2.8 mg/kg (grower) resulted in a significant decrease in performance (a 10% decrease in live weight gain compared to the control groups). DON absorption from the digestive tract was not prevented by MP.

The only consistent effect of MP was to decrease the crude fat digestibility both in the control and in the DON diet. No interactions were detected between wheat and MP addition which indicates a mycotoxin independent effect of MP. Similar unspecific effects on digestibility were also reported for broilers, laying hens and wethers [40-42].

Classical chemical interventions, is the use of aflatoxin-selective clay (HSCAS) as an enterosorbent in the diet to decrease or block the initial exposure of animals to aflatoxins. In early studies, HSCAS clay was reported to rats, chicks, turkey, lambs, and pigs [43]. No nutrient interactions have been reported with HSCAS at the 0.5% level in the diet. HSCAS clay did not impair phytate or inorganic phosphorous utilization. In other studies, the addition at concentrations of 0.5% or 1% did not impair the utilization of riboflavin, vitamin A, or magnese; however there was a slight reduction in zinc utilization in the presence of 1% clay, but not interaction of HSCAS with zinc occurred at 0.5% clay [43,44]. Southern et al. [45] showed that interaction of HSCAS and sodium bentonite clay would affect growth performance and tibia mineral concentrations in chicks fed nutrient-deficient diets. Neither clay affected adversely growth or tibia mineral concentration. The efficacy of a combined supplement of antioxidants (selenium, methionine, and vitamin E) and HSCAS clay was also compared with HSCAS alone or antioxidants individually in 7-day old chickens. A combination of HSCAS and antioxidants provided effective protection against aflatoxin in the diet compared to HSCAS or any antioxidants alone [46].

Copper (Cu) has received considerable attention due to its antimicrobial properties that improve performance in animals when fed over the minimum requirement. Studies have shown that supplementation with copper sources increases growth in poultry and swine [47,48]. Butkeraitis [49] evaluated modified montmorillonites (MONT) in poultry diets. Chick performance was not negatively affected by the addition of 0.5% MONT or 0.5% Cu-MONT to the basal diet, and the absence of any signs of nutritional deficiencies in any of the organs examined indicated that both adsorbents did not negatively affect dietary nutrients (such as minerals and vitamins) at this dietary inclusion level. Addition of Cu-MONT to the AF diet reduced the growth depressing effects. These results suggest that both MONT and Cu-MONT at 0.5% of the diet were partially effective in ameliorating the toxic effects of AF in broilers.

Very promising chemical interventions have been reported and reviewed. These include modified nutrients, as well as non-nutrient dietary additives [50-52]. Alternative methods such as antifungal drugs or adsorbents may also be used with success. Feed manufacturers often utilize binders of various types in feeds to tie up mycotoxins. An alternative for physical and chemical methods of decontamination should be degradation by selected microorganisms, giving a possibility of removal of mycotoxins under mild conditions, without significant losses in nutritive value and the palatability of detoxified food or feed.

New methods include clay-based binders and yeast cell wall extracts and according to Raymond et al. [53] the supplementation with 0.2% yeast cell wall extract (a polymeric glucomannan mycotoxin adsorbent) improved feed intake over feeding contaminated feed alone but not when compared to control diets not containing mycotoxins. The probiotic bacteria and yeast applied creates a starter culture for flour fermentation that has a stable feature of detoxication of aflatoxins and especially ochratoxin A. Using this probiotic starter allows the proper fermentation of a microflora pattern and reduces the risk of development of undesirable and harmful bacteria and moulds. Upon fermentation of hexoses, lactic acid is produced by homofermentation or equimolar amounts of lactic acid, acetic acid/ethanol and CO_2 are produced by heterofermentation. When a mixture of acids is present, it is likely that lactic acid contributes mainly to the reduction in pH, while propionic and acetic acid, which become undissociated, are the actual antimicrobial agents. Acetic acid is the strongest inhibitor and has a wide range of inhibitory activity, inhibiting yeasts, moulds and bacteria, while propionic acid has been observed to exert a strong antifungal effect. Other general antimicrobial substances produced widely by lactic acid bacteria include hydrogen peroxide, carbon dioxide, diacetyl and bacteriocins [54].

In vitro adsorption of zearalenone (ZEA) by cetyltrimethyl ammonium bromide (CTAB)-modified montmorillonite nanocomposites (CMN) was investigated. The particle size of CMN was below 38.06 nm determined with atomic force microscopes. The coexistence of Ca^{2+}, Fe^{3+}, Fe^{2+}, vitamin E and/or lysine showed no significant influence on ZEA adsorption by CMN, suggesting that the nanograde particle size and hydrophobic property of CMN are responsible for the specific adsorption. These results suggest that CMN has the potential as a novel and promising nanograde adsorbent to bind the mycotoxin effectively in the animal gastrointestinal tract to reduce its absorption [55].

The efficacy of melatonin co-administration on aflatoxicosis in chicks was investigated. Melatonin has been shown to protect against oxidative stress in various, highly divergent experimental systems. There are many reasons for its remarkable protective potential. Melatonin is an antioxidant that can easily cross cell membranes and the blood-brain barrier and is considered a direct scavenger of OH, O_2^-, and NO. In the USA, because it is sold as a dietary supplement and not as a drug, the Food and Drug Administration (FDA) regulations that apply to medications are not applicable to melatonin [56-58].

Özen et al. [59] studied in chicks feed aflatoxin plus melatonin. One of the groups was kept as a control (C), and the others were given 150 ppb aflatoxin (AF1), 300 ppb aflatoxin (AF2), 150 ppb aflatoxin plus 10 mg/kg/bwt melatonin (AF1+M), 300 ppb aflatoxin plus 10 mg/kg/bwt melatonin (AF2+M), 10 mg/kg/bwt melatonin (M), and 1% ethanol (E). After 21 day-treatment period, the chicks were sacrificed. Pathological changes were markedly reduced in AF+M groups, and a microscopic view similar to group C was observed. Immunoreactivity in AF+M groups was markedly reduced compared to AF groups and was similar to group C in liver and kidney. Many apoptotic cells were detected in the livers of AF groups, whereas there were no apoptotic cells in AF+M groups. According to the authors these results indicated that tissue degeneration caused by aflatoxin could be greatly reduced by melatonin supplementation in chicks.

Conclusion

There is no general treatment to all mycotoxins and to guarantee the quality of animal feed, good manufacturing practices and monitoring programs to detect the maximum limit of aflatoxins allowed in grain are still important in all steps of grain production. The quality and balance of animal nutrition is an important way to minimize the mycotoxicosis effects on animal health. The long-term effects of binding agents are not clear; therefore, they should be used with caution and for each new strategy to specific mycotoxin decontamination or method prevention, more studies associated to nutritional impacts are necessary, to improve the animal performance and the effects should be examined in more detail since unspecific effects, as an unspecific binding of several substances, bear the danger of an uncontrolled nutrient or micro-nutrient depletion.

References

[1] BottomLey, RA; Christensen, CM; Geddes, WF. Grain storage studies, X. The influence of aeration time and moisture content on fat acidity, non reducing sugars and mould flora of stored yellow corn. *Cereal Chemistry*, 1952; 29: 53-64.

[2] Takigawa, A; Ohyama, Y. Changes of fat in some milled grains during storage. *Bulletin of National Institute of Animal Industry*, 1976; 30: 1-7.

[3] Isaac, S. Mycology answers, *Mycologist*, 1997; 11: 92-93.

[4] Latgé, JP. The pathobiology of *Aspergillus fumigatus*. *Trends in Microbiology*. 2001; 9:8.

[5] Davies, RW. Heterologous gene expression and protein secretion in *Aspergillus*. *Program Industrial Microbiology*. 1994; 29:527–560.

[6] Wyatt, RD; Neathery, MW; Moos, W H; Miller, WJ; Gentry, RP; Ware, GO. Effects of Dietary Aflatoxin and Zinc on Enzymes and Other Blood Constituents in Dairy Calves. *Journal of Dairy Science*. 1985:68;2.

[7] Dänicke, S; Valenta, H; Döll, S; Ganter, M; Flachowsky, G. On the effectiveness of a detoxifying agent in preventing fusario-toxicosis in fattening pigs. *Animal Feed Science and Technology*. 2004; 114: 141–157.

[8] Hamilton, RMG., Trenholm, HL. Chemical, nutritive, deoxynivalenol and zearalenone content of corn relative to the site of inoculation with different isolates of *Fusarium graminearum*. *Journal of the Science of Food and Agriculture*. 1988; 43: 37–47.

[9] Blaha, J; Jicfnska, E; Vesely, D; Jelinek, R. The Effect of Moulds on the Nutritional Value of Wheat. *Animal Feed Science and Technology*. 1990; 28:315-324.

[10] Christensen, CM, Nelson, GH, Speers, GM; Mirocha, CJ. Results of feeding tests with rations containing grain invaded by mixture of naturally present fungi plus *Aspergillus flavus* NRRL 2999. Feedstuffs. 1973; 45: 20-41.

[11] Stringhini, JH; Mogyca, NS; Andrade, MA; Orsine, G F; Café, MB; Borges, SA. Efeito da Qualidade do Milho no Desempenho de Frangos de Corte. *Revista Brasileira de Zootecnia*. 2000:29;1.

[12] Lopez-Garcia, R; Park, DL. Effectiveness of postharvest procedures in management of mycotoxin hazards. In: Sinha, KK & Bhatnagar, D, editors. *Mycotoxins in Agriculture and Food Safety*. Marcel Dekker, Inc., New York, 1998; 407 -433.

[13] Sinha, KK. Detoxification of mycotoxins and food safety. In: Sinha, KK & Bhatnagar, D, editors. *Mycotoxins in Agriculture and Food Safety*. Marcel Dekker, Inc., New York, 1998; 381-405.

[14] Brucato, M, Sundlof, SF, Bell, JU; Edds, GT. Aflatoxin B_1 toxicosis in dairy calves pretreated with selenium - vitamine E. *American Journal of Veterinary Research*. 1986; 47:179 -183.

[15] Galvano, F; Piva, A; Ritieni, A; Galvano, G. Dietary strategies to counteract the effects of mycotoxins: A review. *Journal of Food Protection*. 2001;64:120 –131.

[16] Smith, JW; Hill, CH; Hamilton,PB. The effect of dietary modifications on aflatoxicosis in the broiler chicken. *Poultry Science*. 1971;50:768 -771.

[17] Osborne DJ, Hamilton PB. Decreased pancreatic digestive enzymes during aflatoxicosis. *Poultry Science*. 1981; 60: 1818-1822.

[18] Whiting, F. The feeding value of damaged grain. *Canada Department of Agriculture Publication*, 1954; 1-13.

[19] Sharby, TF; Templeton, GE; Besley, JN; Stephenson, EL. Toxicity resulting from feeding experimentally molded corn to broiler chicks. *Poultry Science*. 1973;52: 1007-1014.

[20] Food and Agriculture Organization. Post-harvest losses in quality of food grains. *Food and Nutrition Paper*. Rome, 1983. 29: 17-44.

[21] Cole, RJ; Kirksey, JW; Dorner, DM; Wilson, JC; Johnson Jr., AN; Johnson, DM; et al. Mycotoxins produced by *Aspergillus fumigatus* species isolated from moldy silage. *Journal of Agricultural and Food Chemistry*. 1977; 25:826 -830.

[22] Morgavi, DP; Boudra, H; Jouany, JP; Michalet-Doreau, B. Effect and stability of gliotoxin, an *Aspergillus fumigatus* toxin, on *in vitro* rumen fermentation. *Food Additives & Containants*. 2004; 21:871 -878.

[23] Newberne PM, Wogan GN, Hall A. Effects of dietary modification on response of the duckling to aflatoxin. *Journal of Nutrition* 1966; 90: 123-26.

[24] Degen GH, Newmann HG. The major metabolite of aflatoxin B_1 in the rat is glutathione conjugate. *Chemico- Biological Interactions.* 1978; 22: 239-241.

[25] Dalvi RR, Ademoyero AA. Toxic effects of aflatoxin B_1 in chickens given feed contaminated with *Aspergillus flavus* and reduction of the toxicity by activated charcoal and some chemical agents. *Avian Diseases.* 1984; 28: 61-68.

[26] Veltmann JR, Wyatt RD, Voigt MN. The effect of varying the total sulfur amino acid content in the diets of chicks with aflatoxicosis. *Poultry Science.* 1981; 60: 1748-51.

[27] Sisk DB; Carlton WW. Effect of dietary protein concentration on response of miniature swine to aflatoxins. *American Journal Veterinary Research.* 1972; 33: 107-14.

[28] Smith TW; Hamilton PB. Aflatoxicosis in the broiler chicken. *Poultry Science.* 1970; 49: 207-15.

[29] Smith TK. Influence of mycotoxins on protein and amino acid utilization. *Federations Proceedings.* 1982; 41 (11): 2828-32.

[30] Verma, J; Johri, TS; Swain, BK. Effect of aflatoxin, ochratoxin and their combination on protein and energy utilization in white leghorn laying hens. *Journal of the Science of Food and Agriculture.* 2007; 87 (5): 760-764.

[31] National Research Council. Nutrient Requirements of Poultry, 9th ed. *National Academy of Sciences*, Washington, D.C. 1994.

[32] Leeson S, Diaz G, Summers JD. Poultry metabolic disorders and mycotoxins. *University Books*, Guelph, Ontario. 1995.

[33] Santurio JM Mycotoxins and Mycotoxicosis in Poultry. *Revista. Brasileira de Ciência Avícola.* 2000; 1: 2.

[34] Bryden, WL, Cumming, RB; Balnave, D. The influence of vitamin A status on the response of chickens to aflatoxin B_1 and changes in liver lipid metabolism associated with aflatoxicosis. *British Journal of Nutrition.* 1979; 41: 529-541.

[35] Cassand P, Decoudu S, Lévêque F, Daubèze M, Narbonne JF. Effect of vitamin E dietary intake on in vitro activation of aflatoxin B_1. Mutation *Research.* 1993; 319(4), 309-316.

[36] Ramos, JJ; Fernández, A; Saez, T; Sanz, MC; Marca, MC. Effect of aflatoxicosis on blood mineral constituents of growing lambs. *Small Ruminant Research.* 1996; 21(3):233-238.

[37] CAST-Council for Agricultural Science and Technology, Task Force Report. Ames, USA, 1989; (116): 1-91.

[38] Madsen, A; Mortensen, HP; Hjarde, W; Leerbeck, E; Leth, T. Vitamin E in barley treated with propionic acid with special reference to the feeding of bacon pigs. *Acta Agriculturae Scandinavica.* 1973; (19):169.

[39] Sekiyama, BL; Ferrari, G; Junior, MM. Processos de descontaminação de rações contendo micotoxinas. *Revista Analytica.* 2007: 26.

[40] Dänicke, S, Ueberschär, KH, Halle, I, Matthes, S, Valenta, H, Flachowsky, G. Effect of addition of a detoxifying agent to laying hen diets containing uncontaminated or *Fusarium* toxin-contaminated maize on performance of hens and on carryover of zearalenone. *Poultry Science.* 2002a; 81: 1671–1680.

[41] Dänicke, S, Gädeken, D, Ueberschär, KH, Meyer, U, Scholz, H. Effects of *Fusarium* toxin contaminated wheat and of a detoxifying agent on performance of growing bulls, on nutrient digestibility in wethers and on the carry over of zearalenone. *Archives of Animal Nutrition*, 2002b; 56: 245–261.

[42] Dänicke, S, Matthes, S, Halle, I, Ueberschär, KH, Döll, S, Valenta, H. Effects of graded levels of *Fusarium* toxin-contaminated wheat and of a detoxifying agent in broiler diets on performance, nutrient digestibility and blood chemical parameters. *British Poultry Science*. 2003; 44: 113–126.

[43] Philips, TD. Dietary Clay in the Chemoprevention of Aflatoxin-Induced Disease. *Toxicological Sciences*. 1999; 52:118-126.

[44] Chung, TK; Erdman, JW; Baker, DH. Hydrated sodium calcium aluminosilicate: Effects on zinc, manganese, vitamin A and riboflavin utilization. *Poultry Science,* 1990; 69:1364-1370.

[45] Southern LL; Ward TL; Bidner TD; Hebert, LG. Effect of sodium bentonite or hydrated sodium calcium aluminosilicate on growth performance and tibia mineral concentrations in broiler chicks fed nutrient deficient diets. *Poultry Science*, 1994;73: 848-854.

[46] Nahm, KH. Prevention of aflatoxicosis by addition of antioxidants and hydrated sodium calcium aluminosilicate to the diet of young chicks. *Japan Poultry Science.*1995; 32:117-127.

[47] Pesti, GM; Bakalli, RI. Studies on the feeding of cupric sulfate pentahydrate and cupric citrate to broiler chickens. *Poultry Science*, 1996; 75:1086-1091.

[48] Hill, GM; Cromwell, GL; Crenshaw, TD; Dove, CR; Ewan, RC; Knabe, DA et al. Growth promotion effects and plasma changes from feeding high dietary concentrations of zinc and copper to weanling pigs (regional study). *Journal of Animal Science.* 2000; 78:1010-1016.

[49] Butkeraitis, P. Evaluation of modified montmorillonites in poultry diets. Dissertation, University of Missouri-Columbia, December, 2007.

[50] Kensler, TW; Davis, EF; Bolton, MG. Strategies for chemoprotection against aflatoxin-induced liver cancer. In: Eaton, LD and Groopman, JD, Editors. *The toxicology of aflatoxin: Human Health, Veterinary and Agricultural Significance*, Academic Press, New York, 1994; 281-306.

[51] Rogers, AE. Nutritional modulation of aflatoxin carcinogenesis. In: Eaton, LD and Groopman, JD, Editors. *The toxicology of aflatoxin: Human Health, Veterinary and Agricultural Significance*, Academic Press, New York, 1994; 207-230.

[52] Yuongman, LD, Campell, TC. Inhibition of aflatoxin B_1 induced gamma-glutamyltranspeptidase positive (GGT+) hepatic preneoplastic foci and tumors by low diets: Evidence that altered GGT+ foci indicate neoplastic potential. *Carcinogenesis.* 1992;13:1607-1613.

[53] Raymond SL, Smith TK, Swamy HVLN. Effects of feeding of grains naturally contaminated with *Fusarium* mycotoxins on feed intake, serum chemistry, and hematology of horses, and the efficacy of a polymeric glucomannan mycotoxin adsorbent. *Journal of Animal Science*. 2003;81: 2123-2130.

[54] Biernasiak, J; Piotrowska, M; Libudzisz, Z. Detoxification of mycotoxins by probiotic preparation for broiler chickens. *Mycotoxin Research.* 2006; 22 (4):230-235.

[55] Feng, J; Shan, M; Du, H; Han X; Xu Z. *In vitro* adsorption of zearalenone by cetyltrimethyl ammonium bromide-modified montmorillonite nanocomposites *Microporous and Mesoporous Materials.* 2008; 113:99-105.

[56] Hardeland R. Antioxidative protection by melatonin: multiplicity of mechanisms from radical detoxification to radical avoidance. *Endocrine.* 2005; 27 (2): 119–130.

[57] Altun A, Ugur-Altun B. Melatonin: therapeutic and clinical utilization. *International Journal of Clinical Practice.* 2007; 61 (5): 835–45.

[58] Poeggeler, B; Saarela, S; Reiter, RJ; Tan, DX; Chen, LD; Manchester, LC et al. Melatonin--a highly potent endogenous radical scavenger and electron donor: new aspects of the oxidation chemistry of this indole accessed in vitro. *Annals of the New York Academy of Sciences*, 1994; 17:419–420.

[59] Özen, H, Karaman, M, Çiğremiş, Y, Tuzcu, M, Özcan, K; Erdağ, D. Effectiveness of melatonin on aflatoxicosis in chicks. *Research in Veterinary Science*. 2009; 86: 485-489.

In: Mycotoxicoses in Animals Economically Important ISBN: 978-1-61668-195-1
Editors: Edlyane Gonçalez, et. al. pp. 143-183

Chapter 7

METHODS FOR PREVENTING AND CONTROL MYCOTOXINS AND MYCOTOXIGENIC FUNGI

Joana D'arc Felicio[1*], Simone Aquino[2] and Edlayne Gonçalez[1]
[1] Centro de Pesquisa e Desenvolvimento de Sanidade Animal, Instituto Biológico, Av. Cons. Rodrigues Alves, 1252, São Paulo, SP, Brazil, CEP 04014-002
[2] Codeagro, Av. Miguel Estéfano, 3.900, São Paulo, SP, Brazil, Cep 04301-903

ABSTRACT

Mycotoxins are metabolites and toxic substances produced by certain filamentous fungi that frequently contaminate food and agriculture commodities, which cause disease in animals or man. A variety of approaches including physical, biological and chemical, treatments to minimize the risk, decontaminate or eliminate the fungi have been attempted. Physical processes used on mycotoxins control involve since manual separation of the contaminated parts until ionizing process. Biological inactivation has been studied using bacteria and fungi to control and inactivation of the mycotoxins production and fungal growth in food and feed. Another way to prevent or interfere with fungal growth and mycotoxin production is by use of natural and synthetic antifungal substances.

INTRODUCTION

According to current estimates, about 100,000 moulds have been identified, from which over 400 can be considered potentially toxic and of which about 5% are known to produce toxic compounds or classes of compounds causing problems in one or more parts of the world [1].

Feeds contaminated with fungal toxins (mycotoxins) pose a health risk to animals and, as a consequence, may cause big economical losses due to the lower efficacy of animal

* Corresponding author: J.D. Felicio
Telephone: +55-11-50871753; Fax: +55-11-50871754
E-mail address: felicio@biologico.sp.gov.br

husbandry. In addition, directly or indirectly (carry through to animal products), contaminated foods may also pose a health risk to humans. So it is understandable that much research has been conducted in an attempt to salvage mycotoxin-contaminated commodities and to avert health risks associated with the toxins [1].

Current control measures are aimed at controlling fungal growth and mycotoxin formation in stored grains by physical methods (aeration, cooling, gamma radiation and modified atmospheres), by chemical methods (ammoniation, acids, bases and food preservatives) or by biological methods. The process of physical, chemical, and biological methods must ensure that the degradation process retains the nutritive value, will not result in the introduction of new toxic or carcinogenic-mutagenic substances, and the process must also destroys fungal spores and mycelia that could, under favorable conditions, proliferate and produce new toxins [2–4].

According with Bata et al. [1], there are three possibilities to avoid harmful effects of contamination of food and feed caused by mycotoxins:

- Prevention of contamination,
- Decontamination of mycotoxin-containing food and feed
- Inhibition of absorption of mycotoxin in consumed food in the digestive tract.

Since contamination of food and feeds with mycotoxins can occurs despite the most strenuous efforts at prevention and since contamination can occur in the field before harvest, the development of the methods that degrade and prevent the mycotoxins formation are very important. This chapter will approach same methods applied to degrade, prevent and control of mycotoxins in food and feeds.

Physical Methods

Physical Separation

The most frequently factor in deterioration of food crops is physical. Physical processes used on mycotoxins control involve separation of the contaminated parts, removal of mycotoxins and inactivation of mycotoxins by heat, cooking, roasting, and radiation are applied on feed [5, 6]. For post-harvest mycotoxin control prevention of conditions that favor fungal growth and subsequent toxin production needs to be considered, as factors such as water activity (*Aw*) of stored products, temperature, grain condition, gas composition of the intergranular air, microbial interactions, and presence of chemical or biological preservatives [7].

The first step to mycotoxin prevention is the physical separation of the mold-damaged kernel, seed or nut from the intact and apparently non-contaminated. Some common physical methods employed are mechanical separation of broken kernels, density segregation, color sorting, and screening. Electronic and hand sorting, density segregation and combinations thereof have been reported for removal of aflatoxin contamination in peanuts [8]. Simple washing procedures using water or sodium carbonate solution resulted also in some reduction of mycotoxins in maize and other grains [9]. However, physical methods of separation, thermal inactivation, irradiation, solvents extraction, adsorption from solution, microbial inactivation, and fermentation are described to prevent the fungal growth or avoid the mycotoxins consumption in exceeded limits. Others physical methods of grain drying include mechanical drying, in-bin drying, infrared, microwave, controlled atmosphere storage or sonic and solar energy drying are applied [10].

Drying, Moisture and Aeration Control

Warm humid climate provides congenial atmosphere for the growth of fungi and production of toxins. Food grains are normally harvested at higher moisture content and then dried to bring down the moisture content up to safe level before storage. From harvesting of the crop till storage conditions may also enhance the chances of microbial invasion/fungal attack at each and every stage/step and production of mycotoxins. *Aspergillus flavus* which is known as storage fungi may infect and produce aflatoxins in crops in the fields also [11]. Delay in drying to safe moisture levels increases risks of mould growth and mycotoxin production. Often harvested maize is left at drying facilities during this critical part of the chain if drying facilities are working at full capacity. This can create problems with an opportunity for growth and mycotoxin contamination of maize, especially by *Fusarium* and *Aspergillus* [12]. It is an established fact that dry grain stores long, safe from insects and moulds because the requirements of moisture for their development are not met. Studies with maize showed that fungal invasion and aflatoxin content could be unacceptable before the grain had lost 0.5% dry matter and mould became visible [13].

Aflatoxins are moderately stable to the roasting process. Dry roasting and oil roasting of groundnut reduces aflatoxins to a significant degree. Dry roasting of naturally contaminated peanuts produced 40 to 50% loss of aflatoxins B_1 and G_1 and 20 to 40% reduction of aflatoxin B_2 and G_2, whereas microwave roasting of contaminated peanuts almost completely destroyed aflatoxins. Previous study demonstrated that microwave roasting (0.7 kW) was more effective than oven roasting in destroying aflatoxins, but higher energy was required (1.6 to 3.2 kW) to achieve 95% destruction, resulting in elevated temperature in peanuts [14]. However, the effect of heating of contaminated feeds in a microwave was found to be the least effective in degradation of aflatoxins comparing to gamma radiation and sunlight. This could be due to the resistance of most mycotoxins to heat [15]. Less information is available on deoxynivalenol. It is stable in many processes and a high pH and temperature are required for breakdown in maize. The decrease during extrusion of maize grits was related to temperature as was found with DON [12]. Experimental roasting corn at 145°C to 165°C reduced aflatoxin B_1 concentration by 40 to 80%. In a study with green coffee beans, heating destroyed 68% of added sterigmatocystin, as well as over 80% of aflatoxin B_1 and ochratoxin A at 200°C for 20 minutes. However, the heat stability of ochratoxin A in green coffee beans inoculated with *Aspergillus alutaceus* was demonstrated. Heat treatment (roasting) at 200 °C for 10 or 20 min reduced the levels of ochratoxin A by only 0–12% in the dried whole beans [16-18]. Cooking at atmospheric pressure can destroy about 50% of the toxins. Cooking rice under 15 Ibs. pressure for 5 minutes degraded 72% of aflatoxins as compared to ordinary cooking or cooking with excess water [11].

Moisture content of grain plays a crucial role in post-harvest processing and is associated with most of the induced characteristics. Water vapour will diffuse throughout a bulk of grain and the moisture will tend to equalize. "Hot spots" may occur at a site of increased respiration (caused by sprouting, infestation or microbial activity), and condensation may occur on cold grain or containers. Different grains have different moisture equilibrium with air, and for each grain type and the level changes slightly with temperature. Typically, however, grain at a safe moisture level of 12 to 13% is in equilibrium with air and with relative humidity (RH) of about 65%. Harvesting of maize is often carried out at moisture contents which are >14–15% which requires drying to reduce the water activity to < 0.70 (=14%) which is safe for storage [12].

Some changes in grain moisture content occur as a result of aeration that changes the temperature of stored grain. Temperature differences in a bin of stored grain cause moisture to migrate from warmer areas to colder areas. Aeration is a process of forcing air through grain to reduce its temperature. It is a very useful storage management tool which can preserve grain from deterioration, especially where the moisture content of the grain is above its safe level. Aeration can be used as effectively in sealed stores as in unsealed ones - sealed stores merely requiring the provision of an air-exhaust ventilator which can be sealed whenever fumigation is to be carried out [10].

Aeration is the practice of moving air through stored grain to prevent storage losses and reduce the grain deterioration, in response to seasonal temperature. It changes, equalizes and maintains uniform temperatures throughout the storage. Aeration is not a grain drying system and must not be considered as such. Spoilage in stored grain is caused by fungal growth, which is related to the moisture content and temperature of the stored grain. Reducing grain temperatures by aeration offers numerous benefits. This method improves the stability of grain during storage by maintaining a cool, uniform temperature, avoids mould development and insect activity and to prevent moisture migration. Besides, it preserves germination viability and it prolongs the effectiveness of insecticide chemicals where these are used. It is apparent that ambient aeration requires periods (e.g. at night time) of low wet bulb temperature to effect cooling of the grain. Such conditions are not always available in tropical climates, particularly during the wet season. The major requirement during such times is more often on drying rather than cooling, however where aeration cooling is required or warranted (for instance after drying, or to delay mould development prior to drying), some air cooling and/or dehumidification may be needed to achieve the requisite conditions [10].

The *Aw* is an influent factor on growth and mycotoxin production. According to Samapundo [19] the use of models to describe the effect of *Aw* and temperature on the radial growth of *Fusarium verticillioides* and *F. proliferatum* on corn can be applied for the prevention of *Fusarium* growth on maize. The growth of these two isolates on corn was therefore studied at water activities between 0.81–0.98 and temperatures between 15 and 30°C. Minimum *Aw* for growth was 0.86 and 0.85 for *F. verticillioides* and *F. proliferatum*, respectively. No growth was observed at *Aw* values equal to 0.83 for both. Previously, Frisvad and Samson [20] estimated that the threshold for growth and OTA production by *Penicillium verrucosum* would be approximately at 0.81–0.83 *Aw* and 0.83–0.90 *Aw*, respectively. The modelling of Lindblad et al.[21] and those of Cairns-Fuller et al.[22] suggest that growth can certainly occur under some conditions at 0.80 *Aw*, although OTA production may be limited to about 0.83 *Aw*. Thus the so-called "zone of uncertainty" which exists for OTA contamination is between 15 and 17.5% moisture content which is critical in determining whether a high risk from OTA can occur. The results confirm the general finding that *Aw* has a greater influence on fungal growth than temperature.

Modified Atmosphere

Modified atmosphere storage has been examined for the storage of moist grain, especially for animal feed. Giorni et al. [23] evaluated the potential for using modified atmospheres of 25–75% CO_2 (balanced with N_2) and *Aw* (0.95, 0.92) to control *Aspergillus flavus* development and aflatoxin B_1 production on maize grain during storage for up to 21 days at 25 °C. The grain inoculated with spores of *A. flavus* there was significantly higher populations of the species at 0.95 than 0.92 of *Aw*. Up to 75% CO_2 resulted in an inhibition of the populations of *A. flavus* isolated from the grain. Aflatoxin B_1 production was obtained and the greatest amounts were produced at 0.95 on maize grain. Overall, the efficacy of controlled atmospheres × *Aw* showed that treatment with 25% CO_2 could be sufficient to efficiently reduce *A. flavus* development but at least 50% CO_2 was required to obtain a significant reduction of aflatoxin synthesis.

Low oxygen concentrations (< 1%) and/or increased concentrations of CO_2 or N_2 have been found to be highly effective in preventing the development of mould on grain and in inhibiting selected mycotoxins, e.g. aflatoxins, ochratoxin, patulin, penicillic acid and T-2 [24]. Studies with *P. verrucosum* and *A. ochraceus* with up to 50% CO_2 suggest that spore germination is not markedly affected, although germ tube extension and hence colonization is significantly inhibited by 50– 75% CO_2, especially at 0.90–0.99 *Aw* for both *P. verrucosum* and *A. ochraceus* [22, 25]. Growth and OTA production were highest in air, followed by 25 and 50% CO_2 regardless of the *Aw* level tested on wheat grain. Generally, CO_2 and *Aw* together cause an enhanced inhibitory effect, although this was not synergistic. More comprehensive studies have been carried out on mycotoxigenic *Aspergillus* and *Fusarium* species. For example, Paster et al. [26] reported that OTA production by *A. ochraceus* was completely inhibited by > 30% CO_2 on agar-based media after 14 days suggesting that there are differences between mycotoxigenic species. This suggests that for efficient storage of moist cereals >50% CO_2 concentrations needs to be achieved rapidly to prevent OTA contamination in storage or during transport.

Paster et al. [27] reported that a lower amount of CO_2 was needed to inhibit fungal development and toxin formation when a reduced O_2 level was applied. The production of zearalenone by an isolate of *Fusarium equiseti* was studied in corn grains stored under modified atmospheres. Zearalenone production was inhibited almost completely in high-moisture corn grains (27%) kept under atmospheres enriched with high CO_2 levels (60%, 40% or 20%) with either 20% or 5% O_2. However, a different result was observed by Wilson and Jay [28] using a low amount of CO_2 on shelled peanuts remoistened to 16.7% and stored for 4 weeks at about 27°C in air (0.03% CO_2, 21% O_2, 78% N_2) and in three modified atmospheres: (1) 13.6% CO_2, 0.3% CO, 0·6% O_2, 84.7% N_2; (2) 12.2% CO_2, 3.1% CO, 0.3% O_2, 83.5% N_2; and (3) 13.6% CO_2, 0.1% CO, 1.5% O_2, 83.9% N_2. The authors reported that the peanuts in air and in the atmosphere containing 1.5% O_2 accumulated high levels of aflatoxin in 1 and 2 weeks, respectively. None of the treatments eliminated infection by the *A. flavus* group. After 4 weeks peanuts in all treatments had visible fungal growth on surfaces.

The levels of CO_2 needed to inhibit mould growth are much higher than those required for the inhibition of mycotoxin production. According to Magan and Aldred [12] elevated CO_2 of >75% are required to ensure that growth of mycotoxigenic moulds does not occur in partially dried grain. The degree of inhibition achieved by elevated CO_2 concentrations is dependent on other environmental factors, such as relative humidity (RH) and temperature. Nevertheless, the biosynthetic pathways for mycotoxin production are merely blocked, but not damaged by high CO_2 levels [24]. All these studies suggest that mycotoxigenic fungi have different responses to exposure to CO_2 in relation to growth and toxin production and factors such as water activity, temperature, nutrients and times of exposure all need to be considered in establishing effective control regimes [10].

Sulphur dioxide (SO_2) is one of the oldest food additives and it is commonly used as a fungal inhibitory treatment. There are problems with corrosion especially of pipework used for delivery of the SO_2 to the grain stores and the presence of SO_2 mainly in the bisulphite form can have moderate toxicity of the solubility products. SO_2 can be highly toxic to microorganisms as it has mutagenic effects and it has deleterious effects on the membrane [29]. In contrast, small quantities of SO_2 may stimulate growth as the sulphur is an essential element for the growth. Some authors found that treatment of 13% moisture content sorghum with sodium metabisulphite inhibited 95% of internal grain fungi, but also significantly reduced germinability [30]. Treatment of moist maize (24%) with 0.3% SO_2 resulted in a significant decrease in microbial colonization and practically no deleterious effect on grain quality. Further studies of moist maize (26.5%) with 0.066% SO_2 alone or combined with 0.018% NH_3 showed that after 60 mould-free days, *Penicillium* spp. were found to have colonized the to third of the bin [31-34]. However, some studies have suggested that much higher concentrations may be required because of adsorption and binding of the SO_2 to the grain, reducing the antifungal activity. Moisture contents of grain (15–19%) may be more effectively treated than moist grain (>20%) as less of the treatment will become bound to the substrate. There is very little detailed information on the actual tolerance and sensitivity of mycotoxigenic fungi to SO_2. Overall, at 0.96 and 0.92 *Aw* up to 2000 mg/L was needed to obtain a 1–2 log decrease in populations of spoilage fungi on cereal grain [10]. Furthermore, Majumber et al. [33] reported the inhibition of both fungal growth and mycotoxin production.

Recent studies with strains of *A. carbonarius* and OTA production suggested that the Effective Dose (ED) 90 values of sodium metabisulphite (mg/ L) used for generating the SO_2 for OTA control were about 650–700 at 0.985 *Aw*, 400–500 at 0.965 *Aw* and 400–450 at 0.93 *Aw*. The difference in sensitivity between spoilage moulds could be due to the fact that several factors influence the efficacy of SO_2. The tolerance of *Penicillium* species to high concentrations of SO_2 has been suggested to be due to their ability to actively transport the SO_2 into the mycelia [34]. Furthermore, the relationship between the concentration of available SO_2 and effects on fungi is influenced by formulation of the product and critically by pH. The solubility products of SO_2 in water vary with the different pH level [29].

Low Temperature Control

Grain itself and the microbial contaminants respire slowly when stored dry. However, if the water availability is increased to 15–19% moisture content (= 0.75–0.85 *Aw*) spoilage fungi, particularly *Aspergillus* and *Penicillium* species grow, resulting in a significant increase in respiratory activity. The chemical process involved in heat generation is predominantly aerobic oxidation of carbohydrates such as starch. This can result in an increase in temperature and sometimes spontaneous heating from the colonization by a succession of fungi resulting in colonization by thermophilic fungi and actinomycetes [35, 36].

Refrigeration is recommended for the reduction of water and respiratory losses during storage. The low temperature does not allow the microflora to grow as most of them are thermophillic [37]. It reduces the metabolic activity and arrest the fungal growth. However, when the microorganisms return to the ambient temperature, fungal growth may commence rapidly. Natural cooling is another effective method of preserving grain, but in the tropics, the use of refrigeration for the storage is limited by economic, managerial and social factors. The association of storage of grain cooling and depleting the oxygen content to the desired level creates a condition where the microbes and insects cannot grow. The heat removed during cooling results in some drying. Grain moisture content will be reduced about one-fourth percent for each 10 °C reduction in temperature. Air-tight storage also works on the same phenomenon where the depletion of oxygen by grain respiration manipulates disinfection by inhibiting aerobic fungi, elimination of mycotoxin production and conservation of desirable quality factors in the grain [37, 38].

Culling Method

Physical separation of infected grains is an efficient and feasible method to minimize mycotoxin contamination. One of the methods used to control or reduce the occurrence of mycotoxin-contaminated nuts and grains is culling and removal of individual contaminated kernels. The mycotoxin content of grains can be further reduced during processing. Automatic colour sorting, often in combination with manual sorting, is widely used to segregate kernels of abnormal appearance (which are considered likely to contain aflatoxin) during the processing of edible grade groundnuts. Mycotoxins can also be concentrated in various fractions produced during the milling process. Zearalenone and deoxynivalenol are reportedly concentrated in the bran fraction during the milling of cereals, however, that all fractions will contain mycotoxins if the original grain is heavily contaminated. Ochratoxin A appears to be reasonably stable to most food processes. In general, the stability of mycotoxins during processing will depend upon a number of factors including grain type, level of contamination, moisture content, temperature and other processing agents [37].

Culling is accomplished by screening during shelling of a crop, removing damage or discolored kernels by hand-sorting on a picking table, and various mechanical and electronic sorters. Mechanical and electronic sorters are used in Brazil and Nigeria. Fungal infection of seeds or grain usually imparts characteristic colour or other physical properties. However, these measures are not completely effective due to existence of "hidden damage", as a fungus contaminates inside the kernel [37-38]. There are machines that separate grains based on fluorescing properties of aflatoxins, as energy reflected from particles illuminated by ultraviolet light. Each type or variety of grain when in optimum health, fully mature, etc. has a characteristic bulk density. This is defined as the weight per standard volume measured in a standard manner. The same characteristic is variously known as "test weight', "bushel weight" or "specific weight". If the bulk density varies the trend is usually downwards and indicative of reduced overall quality of the grain [39]. Pneumatic sorting also is applied to select nuts once moldy nuts are lighter than sound ones and they are sized and conveyed on a sloping screen. When subjected to a partial vacuum, nuts with the lowest density are removed. Another lightweight portion is removed by a blower [14,37,38].

Ionizing and Non Ionizing Radiation

Aflatoxins are affected by exposure to ultraviolet (UV) light; however, practical use of such treatment for destroying aflatoxins is questionable. Feuel [39] reported that exposing a thin layer of peanut meal at a distance of 10 cm below a UV light source for 8 h caused no apparent change in fluorescence of the toxin and kept the toxic effect in ducklings fed with this meal, including death and liver lesions. Light has also been employed successfully to destroy aflatoxin in crude groundnut oil. Studies have shown that visible light is more effective than either ultra-violet or infra-red light [11]. In many countries of the world, the use of solar thermal systems in the agricultural area has shown to be practical, economical and the responsible approach environmentally. Solar heating systems to dry food and other crops can improve the quality of the product, while reducing wasted produce and traditional fuels. Solar food dryers are available in a range of size and design and are used for drying various food products. It is found that various types of driers are available to suit the needs of farmers [40]. The average Indian farmers perform drying of grain conventionally under direct sun light. Olive oil exposed to sun light for 40 minutes caused a 95% loss of aflatoxin. Exposure of aflatoxin contaminated groundnut oil to sun light has given very promising results as it destroyed about 99% of the aflatoxins [41].

Samarajeewa et al. [42] evaluated the detoxification of coconut oil naturally contaminated with aflatoxin B_1. In this study, more than 85% of the toxin is present in the soluble form, the remainder occurring in the sediment. This aflatoxin was detoxified when the oil, in a static layer less than 15 mm thick was exposed to solar radiation of 10 cal/cm^2. At aflatoxin concentrations between 166 and 1250 μg/kg, 75% of the toxin was degraded and total detoxification was achieved on repeated exposure. It was demonstrated that the naturally contaminated coconut oil after exposure to solar radiation did not contain any residual aflatoxins or fluorescent compounds which might have been derived from original aflatoxin B_1.

The ability of ionizing radiation to kill microorganisms has been investigated since the late 19th century and as demonstrated by Aziz and Abd El-Aal [43] the complete elimination of toxigenic moulds in coffee beans and food commodities was achievable with doses from 5 to 10 kGy. The sensitivity of fungi to gamma radiation has been established by many investigators [44-46] who recorded that the dose required for complete inhibition of fungi in different food and feed products ranged from 4 to 6 kGy. There are a number of conflicting reports that show different results in the increase, decrease or even unaffected the production of mycotoxins after irradiation of fungi under various laboratory conditions [47,48]. Irradiation has been shown to destroy the conidia of moulds but the information concerning the effect of irradiation on mycotoxin formation seems to be contradictory. Aflatoxin production was increased in irradiated wheat grain, but decreased in barley and maize when the grain was irradiated prior to inoculation.

However, the dose of radiation depends of constituents, *Aw*, moisture of substrate and fungal burden. The number of spores in the inoculum, grain condition, relative humidity and other environmental factors could all affect the results obtained. However, ochratoxin formation by *Aspergillus alutaceus* was consistently enhanced by irradiation of spores or mycelium [24]. Yun et al. [49] evaluated the effects of compounds as organic acids and amino acids on the radio-degradation of patulin using the dose of 1 kGy. The authors reported that the degradation of patulin was inhibited by the presence of nutritional elements and the radio-degradation effect of patulin is environment dependent. Mitchell [50] showed that the fungal strain, condition of sporage, humidity and irradiation dose affect mould growth and toxin production. There are a number of reports which suggest that moulds are very sensitive to gamma radiation and in addition their mycotoxin production decrease after irradiation.

The direct effect of radiation occurs when the atoms of the DNA molecule are damaged and affect the ability of the microorganism to reproduce and survive, because there is significant alteration in the information carried by the DNA molecule, then the cell may be destroyed by "direct" interference with its life-sustaining system [45]. The presence of water has an important role in the destruction of aflatoxin by gamma energy, since radiolysis of water leads to the formation of highly reactive free radicals. These radicals can readily attack AFB_1 at the terminal furan ring, giving products of lower biological activity [51]. The presence of moisture is an important factors to determinate the effectiveness of irradiation process known as indirect effect.

Aquino et al. [52] evaluated the effect of dose radiation with *Aw* in maize grains. In this study, irradiation process resulted in a reduction of the aflatoxin content of the maize samples comparing the doses of 2, 5 and 10 kGy. The reduction in AFB_2 levels was 97.6% at 2 kGy and 94% at 5 kGy. The same effect was observed with AFB_1 (68.9% at 2 kGy, 46% at 5 kGy). A total degradation of AFB_1 and AFB_2 was obtained with the dose of 10 kGy. The higher sensitivity of AFB_1 and AFB_2 on degradation using the lower dose of 2 kGy was explained by higher *Aw* value at 2 kGy (0.91) compared to 5 kGy (*Aw* 0.88). The gamma radiation associated with high level of *Aw* caused an increased formation of highly reactive free radicals as the radiolytic products of water radiolysis (indirect effect of radiation), but to a total degradation of aflatoxins the dose of 10 kGy was required. The same study showed the decrease of viable cells number of *A. flavus*, with the increasing of doses.

Recently, Aziz and Moussa [47] found that maize, chick-peas and groundnut seeds collected from the Egyptian marked were heavily contaminated by moulds and irradiation at a dose 4.0 kGy reduced mould growth greatly relative to un-irradiated controls and there was no growth of the toxigenic moulds and mycotoxin in seeds using a dose of 5 kGy. The effect of gamma radiation on *Fusarium* moulds and levels of fumonisin B_1 was determined by Aziz et al. [53]. Wheat, maize and barley seeds naturally contaminated with varying levels of fumonisin B_1 1.4–5.8, 8.0–13.8 and 0.1–0.5 μg/g, respectively. *F. moniliforme (verticillioides)* and *Fusarium proliferatum* were major *Fusarium* contaminants producing fumonisin B_1. The viable counts of *Fusarium* in seeds decreased by increasing the radiation dose levels and the growth of *Fusarium* spp. was inhibited at 4 kGy for barley and 6 kGy for wheat and maize. Application of radiation dose at 5 kGy inactivated fumonisin B_1 by 96.6%, 87.1% and 100% for wheat, maize and barley, respectively, and a dose of 7 kGy was sufficient for complete destruction of fumonisin B_1 in wheat and maize.

Ferreira-Castro et al. [54] observed that gamma radiation doses above 5 kGy effectively inhibit growth of *Fusarium verticillioides* in maize grains, although a complete elimination of the fungal microflora required 10 kGy. The fluorescent viability test of spores showed that the samples irradiated with 2 kGy remained viable (36%) and the production of fumonisins increased after irradiation, even along with a reduction in fungal counting by three logarithmic cycles comparing with the control group. The concentrations in the samples irradiated to 2 kGy were higher than the concentrations in the control sample, which may be attributed to the size of the inoculum. The size of the inoculum affects toxin formation, as demonstrated with *A. parasiticus* [55] and *A. flavus* [56]. A medium inoculated with a larger number of spores will develop less aflatoxin than a medium inoculated with a smaller number of spores. The toxin production is apparently suppressed when the number of inoculated spores per unit volume of substrate exceeds a certain level [57]. This result demonstrated that there is no direct correlation between the decrease in fungal counting (cfu/g) and the presence of mycotoxin. The dose of 10 kGy was necessary for eliminating mycotoxins or decreasing their concentrations to an acceptable level, as well as for killing mould. The effects of electron beans and X-rays also been studied and the ionizing radiation is more promise in controlling fungal growth than in destroying mycotoxins.

Herzallah et al. [15] conclude that the energy of sunlight might be effective in destroying aflatoxins than ionizing radiation, in some circumstances. The efficiency of decontamination of aflatoxin residues in poultry feeds through exposure to sunlight (solar radiation), gamma radiation and microwave heating were investigated in artificially contaminated feed samples. Photodegradation of aflatoxin by sunlight has been found to cause a significant ($P < 0.05$) decrease aflatoxins, more than 60% of the aflatoxin was found to be degraded after 30 h of exposure to sunlight. Also, the results show that the treatment of chick feed with the gamma radiation degraded more than 30% of AFB_1 and total aflatoxins in the exposed feed samples to a dose of 25 kGy. The results show that microwave heating is less effective in decontamination of aflatoxins. The percentages of reduction were between 22 to 32%. The extent of detoxification with microwave heating was found to be less effective as a method of decontamination under these conditions with a maximum exposure time of 10 min. The authors concluded that solar decontamination appears to be an efficient and cheaper method to use compared with microwave heating or gamma radiation, but is likely to be impractical on a large commercial scale.

Biological Methods

Many specialists are of the opinion that the best approach for mycotoxin decontamination should be degradation by selected microorganisms, giving a possibility of removal of mycotoxins under mild conditions, without using harmful chemicals and without significant losses in nutritive value and the palatability of detoxified food or feed [1].

A variety of bacteria, yeast and moulds are able to degrade or remove mycotoxins from foods and also can restrict fungal growth. This approach is defined as biodegradation or biotransformation that leads to less toxic products [58].

Aflatoxins

Since the discovery of aflatoxins in the 1960s, researchers have screened numerous natural products and synthetic compounds, as well as extracts from diverse organisms for inhibitors of aflatoxins biosynthesis [59].

Biodegradation of aflatoxins, using microorganisms or enzymes, is one of the well-known strategies for the management of aflatoxins in foods and feeds. The methods of biodegradation are being actively studied and can be a highly promising choice, since it is efficient, specific, and environmentally friendly to reduce or eliminate the possible contaminations of aflatoxins in foods and feeds [60]. Different organisms, including bacteria, yeasts and nontoxigenic *Aspergillus* fungi, have been tested for their ability in the control of aflatoxin contamination [61].

Several species of bacteria from soil, air, and other substrates belong to generous *Bacillus*, *Pseudomonas*, *Burkholderia*, *Stenotrophomonas*, *Agrobacterium*, *Nocardia*, *Lactobacillus*, *Bifidobacterium*, *Lactococcus*, *Rhodococcus*, *Stenotrophomonas*, *Propionibacterium*, *Corynebacterium*, *Mycobacterium*, *Nannocystis* were studied against aflatoxigenic fungi and aflatoxin production [62-77]. These bacteria can degrade aflatoxins, inclusive aflatoxin M_1 [65; 78], with efficiency to less or nontoxic products [60].

The first microorganism with the ability to detoxify aflatoxins was *Flavobacterium aurantiacum* [58], a kind bacterium from soil and water [60].

Lactic acid bacteria have been reported to bind aflatoxins and remove them from food and feeds [77]. Strains of *Lactobacillus*, *Bifidobacterium*, *Propionibacterium* and *Lactococcus* have been reported to degrade aflatoxins [62-65; 67; 71]. The aflatoxin B1 reduction by these bacteria was proven to be mainly by cell binding rather than metabolism or degradation. Most importantly, this kind of binding seems to be reversible, which means that AFB_1 can hardly be removed completely from contaminated media [75].

Lactobacillus casei, *L. plantarum* and *L. fermentum* reduced aflatoxin B_1 (AFB_1) from 25 to 61% at experimental conditions [77]. Khanafari et al. [73] related that *L. plantarum* was incubated with aflatoxin B_1 and after 1h 45% of the toxin was removed and 100% was removed after 90h.

Some saprophytic yeast species (such as *Candida krusei* and *Pichia anomala*) have shown promise as biocontrol agents against *A. flavus*. Similar to bacterial agents, these yeast strains were able to inhibit *Aspergillus* growth greatly in laboratory conditions [61].

Fungi cannot only produce aflatoxins, but they are able to degrade them as well. Fungal strains, *Aspergillus niger*, *Eurotium herbariorum*, *Rhizopus* spp. were tested to remove aflatoxin in experimental conditions [60]. Isolates of *Kluyveromyces* spp. may have practical value in the postharvest *Aspergillus* section Flavi strains control of storage maize [79].

Several fungal cultures were found to prevent synthesis of aflatoxin B_1 in liquid culture medium. Among these *Phoma* spp., *Mucor* spp., *Trichoderma harzianum*, *Trichoderma* spp. 639, *Rhizopus* spp. 663, *Rhizopus* spp. 710, *Rhizopus* spp. 668, *Alternaria* spp. and some strains belonging to the *Sporotrichum* group could inhibit aflatoxin synthesis by > or =90% [80].

Another approach to aflatoxin control involves competitive exclusion of toxigenic *A. flavus* strains with a non-aflatoxigenic *A. flavus* isolate [81]. Great successes in reducing aflatoxin contamination have been achieved by application of non-toxigenic strains of *A. flavus* and *A. parasiticus* in fields of cotton, peanut, maize and pistachio [61; 82-86].

Atehnkeng et al. [85] evaluated *A. flavus* atoxigenic in corn field, and showed that has its potential value as agents for biocontrol of aflatoxin contamination in maize. Another study evaluate the effect of applying non-toxigenic strains of *A. flavus* and *A. parasiticus* to soil separately and in combination on preharvest aflatoxin contamination of peanuts. The authors concluded that treatment with the non-toxigenic *A. flavus* strain alone is more effective than the *A. parasiticus* strain alone and equally as effective as the mixture [84]. Dorner [86], related the treatment of corn field with *A. flavus* atoxigenic, showed that applications of the nontoxigenic strain influenced untreated corn, thus reducing the apparent effect of the biocontrol treatments.

Ochratoxins

Ochratoxins are mycotoxins which exhibit nephrotoxic, immunosuppressive, teratogenic and carcinogenic proprieties [87], and the ochratoxin A is the most potent of them.

Bacillus licheniformis strain was able to inhibit growth *Aspergillus* strains and to remove 92.5% of ochratoxin A (OTA) [88].

Lactic acid bacteria of the *Lactobacillus* and *Lactococcus* genera were assessing their sensitivity to ochratoxin A and their ability to remove it from a liquid medium. It was demonstrated that all 29 strains caused a reduction of the toxin amount in the liquid medium. The highest decrease, exceeding 50% of the initial OTA content, was caused by the strains *Lactobacillus acidophilus* CH-5, *L. rhamnosus* GG, *L. plantarum* BS, *L. brevis* and *L. sanfranciscensis* [89]. Fuchs et al. [90] demonstrated that *Lactobacillus acidophilus* VM 20 caused a decrease of OTA by > or = 95% either.

Grapes contamination with ochratoxin A is due to the attack by ochratoxigenic *Aspergillus carbonarius* and *Aspergillus* section *Nigri*. Four *A. pullulans* strains, AU14-3-1, AU18-3B, AU34-2, and LS30, are resistant to and actively degrade ochratoxin A in vitro. Three of the four strains significantly prevented infections by *A. carbonarius*. Two of these strains were shown to degrade OTA to ochratoxin alpha in fresh grape [91].

Filamentous fungi isolated from Portuguese grapes were assessed for ochratoxin A degradation capabilities. It was observed that 51 of the 76 tested strains, predominantly *Aspergillus* species, were able to degrade more than 80% of ochratoxin A added to the culture medium and that the most potent species (more than 95% of initial amount) were the black aspergilli, *A. clavatus*, *A. alutaceus*, *A. versicolor*, and *A. wentii*. Other fungi frequently isolated from grapes, such as *Alternaria*, *Botrytis*, *Cladosporium*, and *Penicillium*, also showed significant degradation capabilities [92]. *A. fumigatus* and black *A. niger* were found to detoxify ochratoxin A in culture media [93].

Ochratoxin A was successfully degraded by *Rhizopus stolonifer*, *R. microsporus*, *R. homothallicus* and two *R. oryzae* isolates, and by four unidentified *Rhizopus* isolates. *Rhizopus* isolates were able to degrade more than 95% of ochratoxin A within 16 days [91].

A total of 144 yeast isolates were tested in a preliminary screening on agar to select isolates showing a killer activity against *Aspergillus carbonarius* and *A. niger*. Six yeast isolates belonging to five species, namely 2 isolates of *Issatchenkia orientalis* and one each of *Metschnikowia pulcherrima*, *Kluyveromyces thermotolerans*, *Issatchenkia terricola* and *Candida incommunis*, were finally selected and screened on wounded grape berries for their ability to inhibit infection by ochratoxigenic moulds. With the exception of the *K.*

thermotolerans isolate, when inoculated at 10^9 CFU/wound, the other five challenger yeasts reduced the *A. carbonarius* and *A. niger* colonization on grape berry ($P<0.05$) [95].

Patulin

Contamination of apples (*Malus domestica*) and derived juices with fungicide residues and patulin produced by *Penicillium expansum* are major issues of food safety. Patulin has previously been found to be degraded by several yeasts [96]. Biocontrol agents represent an alternative or supplement to chemicals for disease control. Three biocontrol agents, *Rhodotorula glutinis* LS11, *Cryptococcus laurentii* LS28, and *Aureobasidium pullulans* LS30, were examined for their in vitro growth in the presence of patulin and for their capability to decrease mycotoxin recovery from the medium. Strain LS11 yielded the highest growth rates and the greatest decrease of toxin recoveries. These data suggest that biocontrol yeast cells surviving in decaying apples could metabolize patulin and/or negatively affect its accumulation or synthesis [97].

Morales et al. [98] studied *Candida sake* CPA-2 and *Pantoea agglomerans* CPA-1 on *P. expansum* growth and patulin accumulation in cold storage and further deck (ambient) storage. Both microorganisms tested controlled blue rot and patulin accumulation during cold storage. The control of *P. expansum* growth was enhanced in *C. sake* treated apples, on the other side; control of patulin accumulation in *P. agglomerans* treated apples seemed to be more efficient.

A number of *Rhizopus* isolates were able to degrade patulin and it was also degraded by *A. niger* and *A. fumigatus* isolates [94].

Bacteria also are able to degrade patulin, and it was demonstrated by *Bifidobacterium animalis* VM 12, that reduced the toxin levels by 80% [90].

Fumonisins

Fusarium verticillioides is an important plant pathogen, and it also produces the fumonisin mycotoxins, especially fumonisin B_1, which is toxic to livestock and poultry. Furthermore, *F. verticillioides*-contaminated maize is associated with human esophageal cancer and has been recently implicated in human neural tube defects [99].

The genus *Bacillus* is known for the production of over 24 antibiotics, several of which are fungicidal with potential control of plant diseases [99]. *Bacillus mojavensis* was patented and it produced Leu(7)-surfactin, which was toxic to *F. verticillioides* [99]. The bacterium, a strain from a subgroup of *Bacillus subtilis* (patent number 5,994,117), operates under the principle of competitive exclusion, reducing the fumonisin concentration *in planta* [100]. *B. subtilis* CE1 inhibited *F. verticillioides* growth and fumonisin B_1 accumulation in vitro was assayed [101].

Propionibacteria cultures (*Propionibacterium freudenreichii* ssp.) were tested for their ability to inhibit the growth and mycotoxin production of *F. verticillioides*, *F. graminearum*, *F. culmorum* and *Alternaria alternata*. The strongest inhibition of growth and toxin production was observed in the presence of cultures containing viable cells and supernatants obtained from propionibacteria cultures [102].

Bacterial isolates from Mississippi maize field soil and maize rhizosphere samples were evaluated for their potential as biological control agents against *Aspergillus flavus* and *Fusarium verticillioides*. It was identified 221 strains that inhibited growth of both fungi. The most prevalent genera isolated from rhizosphere samples were *Burkholderia* and *Pseudomonas*, and the most prevalent genera isolated from nonrhizosphere soil were *Pseudomonas* and *Bacillus* [74].

Cavaglieri et al. [103] investigated the potential for the biological control of *F. verticillioides* by *Enterobacter cloacae*, *Microbacterium oleovorans*, *Pseudomonas solanacearum* and *Bacillus subtilis*, and the combination *E. cloacae* and *M. oleovorans* has the best results.

Eleven competitive rhizobacterial isolates (*Arthrobacter globiformis* RC1, *Azotobacter armeniacus* RC2, *A. armeniacus* RC3, *A. globiformis* RC4, *A. globiformis* RC5, *A. armeniacus* RC6, *Pseudomonas solanacearum* RC7, *Bacillus subtilis* RC8, B. subtilis RC9, *P. solanacearum* RC10, *B. subtilis* RC11) were tested in against *F. verticillioides*. All bacterial antagonists reduced fumonisin B_1 production at all levels tested [104].

In addition to *B. subtilis*, an isolate of a species of the fungus *Trichoderma* sp. is being tested for postharvest control of fumonisin production in kernels during storage [100]. Culture filtrates of *Trichoderma viride* and *Trichoderma harzianum* were inhibitory of *Fusarium verticillioides* [102]. *T. viride* can suppress FB_1 production by *F. verticillioides*, thereby functioning to control mycotoxin production and the fungal growth [106].

Decontamination by fermentation was evaluated using three yeast strains (*Saccharomyces*), a maximal decrease was observed in the case of 28% and 17% for fumonisin B_1 and fumonisin B2, respectively, and ochratoxin-A 13% [1].

Trichothecenes

The trichothecenes are a group of toxic secondary metabolites produced by several genera of the Fungi Imperfecti, including *Fusarium*, *Myrothecium*, *Trichothecium*, and *Stachybotrys*, some of which naturally contaminate cereal crops during growth or storage in high-moisture environments. Ingestion of moldy crops contaminated in this way has resulted in a number of serious outbreaks of mycotoxicoses in both humans and domesticated animals [107].

The ability of *Lactobacillus* and *Propionibacterium* strain to remove common *Fusarium* toxins, trichothecenes (deoxynivalenol, 3-acetyldeoxynivalenol, nivalenol, fusarenon, diacetoxyscirpenol, T-2 toxin and HT-2 toxin), from liquid media was studied. *Lactobacillus rhamnosus* strain GG and *Propionibacterium freudenreichii* ssp. *shermanii* JS were more efficient than *Lactobacillus rhamnosus* strain LC-705 in removing the toxins from the liquid media [108].

Fusarium graminearum (perfect state, *Gibberella zeae*) infects wheat heads at flowering, causing the disease *Fusarium* head blight (FHB). The pathogen often produces deoxynivalenol resulting in grain size and quality reduction [109]. Deoxynivalenol is a trichothecene secondary metabolite produced by *Fusarium* species infecting cereal crops. As a mycotoxin, DON causes losses in livestock production and poses a health risk to humans consuming contaminated cereal products [110]. Microbial isolates obtained from wheat anthers were screened for their ability to utilize tartaric acid, a compound that is poorly utilized by *F. graminearum* and could be utilized in formulations of biological control agents. Four strains that utilized tartaric acid and three that did not were effective in reducing FHB disease severity by up to 95% in greenhouse and 56% in field trials [109].

A mixed microbial culture capable of metabolizing deoxynivalenol was obtained from soil samples by an enrichment culture procedure. strain E3-39 was classified as a bacterium belonging to the *Agrobacterium-Rhizobium* group removed exogenously supplied deoxynivalenol from culture medium after incubation for 1 day and it 39 was also capable of transforming 3-acetyldeoxynivalenol but not nivalenol and fusarenon-X [107].

Bacterial communities isolated from 17 of 20 samples of soils and waters with widely diverse geographical origins utilized T-2 toxin as a sole source of carbon and energy for growth. These isolates readily detoxified T-2 toxin as assessed by a *Rhodotorula rubra* [111].

Lactobacillus paracasei subsp. *tolerans*, isolated from a traditional sourdough bread culture and previously shown to have antifungal activity against *Fusarium proliferatum* M 5991 and M 5689 and *F. graminearum* R 4053 in a liquid medium [112].

Addition of *Bacillus coagulans* to *Fusarium culmorum* culture inhibits the DON (deoxynivalenol) production. The strongest inhibition of fungal growth was noticed when *Bacillus coagulans* was co-inoculated at the beginning of culture. Estimation of ergosterol level as a determinant of fungal growth showed the greatest degree of *Fusarium* sp. Inhibition [113].

Kluyveromyces marxianus was sensitive to all trichothecene toxins, but this yeast was not inhibited by other mycotoxins [1].

Zearalenone

Zearalenone (ZEA) is a mycotoxin produced mainly by fungi belonging to the genus *Fusarium* in foods and feeds. It is frequently implicated in reproductive disorders of farm animals and occasionally in hyperoestrogenic syndromes in humans. There is evidence that ZEA and its metabolites possess oestrogenic activity in pigs, cattle and sheep [114].

Yeast (*S. cereviseae*), fungi (*Mucor bainieri*, *Streptomyces rimosusand*, *Rhizopus* spp.) and bacterias (*Lactobacillus* strains) can be able to degrade zearalenone [114]. But, a variety of microorganisms were found to be able to convert zearalenone to a- and h-zearalenol. However, this transformation cannot be regarded as detoxification since the oestrogenic activity of these metabolites is similar to that of zearalenone [96].

The propionibacteria strains were tested for their ability to inhibit the growth and mycotoxin production *of Alternaria alternata*, *Fusarium culmorum*, *Fusarium graminearum* and *Fusarium verticillioides*. The strongest inhibition of growth and toxin production was observed in the presence of cultures containing viable cells and supernatants obtained from propionibacteria cultures. The ability of *P. freudenreichii* ssp. *freudenreichii* 111 to remove zearalenone from liquid medium was also evaluated. It was shown that both viable and non-viable cells caused a decrease in zearalenone concentration in the medium [102].

Zearalenone was found to be completely degraded by several *Rhizopus* isolates including *R. stolonifer*, *R. oryzae* and *R. microsporus* strains [94].

CHEMICAL METHODS

Contamination of food commodities with mycotoxins resulting from fungal attack can occur before, after and during the harvest and storage operations. We have been related the state of the art of the use of chemical products for the prevention and control of mycotoxigenic fungi and their mycotoxins. The chemical control involves the use of synthetic; natural substances and extract or essential oil from plants.

Synthetic Chemical

A number of chemicals substances, such as acids, alkalis, aldehydes, oxidizing agents, gases and guanidine-based polymeric compounds have been used on chemical detoxification of feeds (Table 1). However, many such agents are not practical because they may render the produce unsafe. The only economically feasible yet effective decontamination of cottonseed meal was obtained by ammoniation.

Currently ammonia vapors are used to inactivate mycotoxins [115], ammoniation of agricultural products destined for cattle and poultry has been approved to be safe in accordance to the criteria defined by government bodies [116]. Ammoniated corn feed was found to be devoid of any toxic effects in chickens or rats, with a sufficient reduction of aflatoxins [117].

One of the characteristics of aflatoxin deactivation processes is that it should destroy the mycelia and spores of the toxic fungi, which may proliferate under favorable conditions [115]. The ammoniation process using either ammonium hydroxide or gaseous ammonia is recognized as an efficient procedure to reduce aflatoxins levels in animal feeds. When applied to fumonisin-contaminated maize, ammoniation for 4 days at 50 °C and atmospheric pressure decreased fumonisins by 30–45% [118].

The calcium hidroxide $Ca(OH)_2$, is not really effective since it produces hydrolyzed FB_1 which is toxic. But the combination of heat treatment with sodium bicarbonate $NaHCO_3$ and hydrogen dioxide H_2O_2 alone or with $Ca(OH)_2$ gave a complete reduction (100%) of FB_1 in contaminated maize (100mg/kg) [119].

Treatment of contaminated maize with sodium bisulfite solutions (4.4 mg/kg) at 80 °C for 18 h can convert 85% of DON into a DON-sulfonate conjugate which appears to be non toxic to pigs [120]. Other chemical treatments with hydrochloric acid, hydrogen peroxide, sodium hypochlorite, ascorbic acid and ammonium carbonate, did not show any effect against DON. The use of insecticides carbamate-type, insecticide, Decis and similarly showed a strong reduction of Ochratoxin [121]. In general, fungicides that contained copper or strobilurins reduced both growth and OTA production, contrary to sulfur fungicides [122].

Other results using *Aspergillus carbonarius* as *in vitro* control were achieved with several fungicides as penconazole 10%, Fludioxonil50%, Cyprodinil 37.5% plus fludioxonil 25% , Cyprodinil 50% , Folpet 40% plus mefenoxam 5%, Tolyfluanid 50% , Tebuconazole 25% Captan 50%, Carbendazim 50, MANCOZEB 80–1, Tiram 80% , oxychloride 22% plus mancozeb 17.5.% [123].

Barberis (2009) [124] determined the efficacy of phenolic antioxidant butylated hydroxyanisole (BHA) under different interacting with water activity (*Aw*) and temperature regimes on the lag phase and growth rate by *Aspergillus* section *Nigri* strains. At all *Aw* levels tested, BHA at 20 mmol/ L completely inhibited growth and OTA production.

Table 1: Synthetic substances with antifungal activity on toxigenic moulds.

SUBSTANCES	MOULD	TOXIN	REFERENCE NUMBER
2-deoxy-d-glucose	*Penicillium expansum*	Patulin	172
Akacid plus	*Aspergillus parasiticus*		173
Benzoic acid	*Aspergillus niger* *Aspergillus flavus* *Aspergillus alutaceus* *Aspergillus parasiticus*		126
Butylated hydroxanilase (BHT), propil paraben (BHA-PP)	*Aspergillus flavus* *Aspergillus niger*	AFB_1	174
Butylated hydroxytoluene (BHT), propil paraben PP (BHT-PP)	*Aspergillus* section *flavi*	AFB_1	174
Carbendazin	*Aspergillus carbonarius* *Aspergillus* spp.	OTA AFB_1	175 126
Citric acid		AFB_1	127
Imazalil	*Penicillium expansum* *Penicillium* spp.		176
Propiconazole	*Aspergillus* spp.	AFB_1	127
Salicylic acid	*Fusarium oxysporum*		128
Trans – 2 - hexenal	*Aspergillus flavus* *Penicillium expansum*	Patulin	177, 178
Vanillin		AFB_1	126

Systemic benzimidazole fungicides, benomyl, methyl 2-benzimidazole carbamate, and TBZ, were introduced into packinghouses in the 1970s to control postharvest decay caused by *Penicillium expansum* and *Botrytis cinerea*. However, resistant strains to these fungicides developed soon. Development of resistance to commonly used fungicides within populations of postharvest pathogens has become a matter of much concern in recent years [122]. Currently, imazalil is the most common used fungicide applied postharvest, as well as mixtures of imazalil plus folpet or other fungicide products such as iprodione and TBZ [122]. A study *in vitro* have been showed that carbendazim, captan, and bupirimate inhibited patulin production by *P. expansum*, but to a smaller extent than its growth [125]. Recent works related study of effects of Benzoic, salicylic and citric acids on *A. flavus, A. niger, A. ochraceus, A. parasiticus, Fusarium oxysporum* [126, 127, 128].

Synthetic chemicals such as fungicides or preservatives have been used for a long time and have greatly contributed in management of such losses. The application of such chemicals has led to a number of environmental and health problems due to their residual toxicity, carcinogenicity, hormonal imbalance and spermatotoxicity [129, 130].

Substances and Extract or Essential Oil From Plants

Because of indiscriminate use, some microorganisms have developed resistance to most widely used synthetic fungitoxicants rendering them out of date [131]. Hence, there is a need to develop new fungicides or preservatives with improved performance as well as ecofriendly in nature.

During recent years, substantial interest has developed in the preservation of grains by the use of essential oils to effectively retard mycotoxin production as well as the fungal growth. Different extracts and essential oils of spices, herbs and other plant materials are becoming increasingly important in the food industry because of their antifungal.

The increased demand for safe and organic food, without chemical preservatives, provokes many researchers to investigate the antimicrobial effects of natural compounds. Numerous investigations have confirmed the antimicrobial action of essential oils in model food systems and in real food. There is evidence that essential oils are more strongly antimicrobial than is accounted for by the additive effect of their major antimicrobial components; minor components appear, therefore, to play a significant role. Since essential oils are considered as generally regarded as safe, the possibility of reinforcing their natural antimicrobial effects by the addition of small amounts of other natural preservatives may be a way to attain a balance between sensory acceptability and antimicrobial efficacy [132].

Ghorbanian 2007 [133] studied the effects of different concentrations of aqueous neem leaf extract on fungal growth and aflatoxin production by *A. parasiticus* (NRRL 2999) at different incubation times. The toxigenic fungus was cultured on sucrose low salts medium in the presence of various concentrations of extracts (0.2, 0.8, 3.12, 12.5 and 50% v/v). The extracts did not have any obvious effect on fungal growth. AFB_1 production in the control samples increased to the maximum level on the 8th day. The inhibition of aflatoxin synthesis by plant extracts was found to be time - and dose-dependent. The maximum inhibitory effect was 80–90% in the presence of 50% concentration.

The ethanolic extract of *Polymmia sonchifolia* at 200 µg/mL and 250 µg/mL reduced the aflatoxin B_1 and B_2 productions at 84%, 62%, 96% and 98% respectively. The extract was fractioned and hexanic fraction inhibited AFB_1 and fungal growth at 100 µg/mL at 97% and 38% respectively and 99% of the AFB_2 production. Therefore the ethyl acetate fraction inhibited the fungal growth at 100 µg/mL at 34% and aflatoxin production at 87% [134]. The aqueous extract also inhibited AFB_1 and fungal growth [135]. The flavonoids 3', 5, 7 trihydroxy 3, 4'-dimcthoxyflavonc (1) and 3', 4', 5- trihydroxy-7-methoxy flavanone (2) and two sesquiterpenes lactones enhydrin (3) and a mixture of enhydrin and uvedalin (4) isolated from *Polymnia sonchifolia* leaves, were evaluated on the aflatoxin production by *A. flavus*. Compound 1 inhibited 25% of the aflatoxin B_1 production at the concentration of 15 µg/mL and the compound 2 inhibited 8% of the aflatoxin B_1 production the compound 4 inhibited 34% and 76% of the fungal growth and AFB_1 production respectively [136].

The essential oils of plant species mainly of the species of family Labiateae, Compositeae, Lamiaceae, Apiaceae, Asteraceae, Lauraceae, Umbellifereae were related as antifungal on toxigenic fungi (Table 2).

Table 2: Extract and essential oil from plant species with antifungal activity and inhibition of mycotoxins production.

PLANTS	EXTRACT	MOLDS	TOXIN	REFERENCE NUMBER
Adenocalymma alliaceum	Extract	*A. flavus* *A. niger*	AFB_1	187, 188
Ageratum conyzoides	Essential oil	*A. flavus*	AFB_1	139, 151
Allium sativum	Essential oil	*A. flavus*	AFB_1	126
Allium sativum		*A. flavus*		141
Aloysia triphylla		*A. flavus* *A. alutaceus*		189
Azadirachta indica	Essential oil Aqueous extract	*A. parasiticus*	Aflatoxins	133, 44
Baccharis articulata		*A. parasiticus* *A. alutaceus*		189
Baldea boldus		*A. alutaceus* *F. oxysporum*		189
Carum carvi	Essential oil	*A. parasiticus* *A. flavus* *A. alutaceus* *F. verticillioides*	AFB_1 AFG_1 Ochratoxin A fumonisin	144, 190
Cecropia adenopus		*A. parasiticus* *A. alutaceus*		189
Chenopodium ambrosioides	Essential oil	*A. flavus* *A. niger* *A. fumigatus* *F. oxysporum*	AFB_1	130
Cinnamomum zeylanicum	Essential oil	*A. flavus* *A. niger* *A. terreus* *F. verticillioides* *F. graminearum* *P. citrinum* *P. viridicatum*	AFB_1 Zearalenone DON	142, 150, 192,
Citrus aurantifolia	Essential oil	*A. parasiticus*	AFB_1 AFG_1	144, 148
Citrus lemon	Essential oil	*A. niger* *A. flavus* *P. verrucosum* *P. chrysogenum*		140

Table 2. (Continued)

PLANTS	EXTRACT	MOLDS	TOXIN	REFERENCE NUMBER
Citrus paradisi	Essential oil	*A. niger* *A. flavus* *P. verrucosum* *P. chrysogenum*		140
Citrus reticulata	Essential oil	*A. niger* *A. flavus* *P. verrucosum* *P. chrysogenum*		140
Citrus sinensis	Essential oil	*A. niger* *A. flavus* *P. verrucosum* *P. chrysogenum*		140
Curcuma longa	Essential oil	*A. flavus*	AFB_1	126
Cymbopogon citratus	Essential oil Extract	*A. alutaceus* *P. verrucosum* *P. expansum* *A. flavus* *A. niger* *A. flavus* *A. fumigatus* *F. verticillioides*	AFB_1 AFB_2 ,	139, 143,144, 193, 194, 195
Cymbopogon flexuosus	Essential oil	*A. flavus*	AFB_1	194
Cymbopogon winterianus	Essential oil	*A. flavus*	AFB_1	196
Equisetum giganteum		*A. flavus* *A. alutaceus* *F. oxysporum* *F. verticillioides*		189
Eucalyptus globulus	Essential oil	*A. flavus* *A. parasiticus*	AFB_1	153
Foenicculum vulgare	Essential oil	*A. parasiticus* *A. flavus* *A. alutaceus* *F. verticillioides*	Aflatoxins Ochratoxin A fumonisin	190
Hedeoma multiflora	Essential oil	*A. flavus*	AFB_1	145

PLANTS	EXTRACT	MOLDS	TOXIN	REFERENCE NUMBER
Lippia turbinata	Essential oil	*A. flavus* *A. parasiticus* *F. verticillioides*	AFB_1 AFB_2, AFG_1, AFG_2 Fumonisin B_1	126, 145, 197,198
Madhuca indica	Essential oil	*A. flavus*	AFB_1	199
Melissa officinalis		*A. flavus* *F. verticillioides*		189
Menta viridicum	Essential oil	*A. parasiticus* *A. flavus* *A. alutaceus* *F. verticillioides*	Aflatoxins Ochratoxin A fumonisin	190
Occamin ternuiflorum	Extract	*A. flavus*	AFB_1	143
Ocimum basilicum	Essential oil	*A. parasiticus* *A. flavus* *A. alutaceus* *F. verticillioides*	Aflatoxins Ochratoxin A fumonisin	190
Ocimum gratissimum	Essential oil	*A. alutaceus* *P. expansum* *P. verrucosum* *A. flavus* *A. fumigatus* *F. verticillioides*		143, 144
Origanum vulgare	Essential oil	*A. flavus* *F. verticillioides*	AFB_1 FB_1	192, 200,201,202
Passiflora coerulea		*A. flavus* *A. alutaceus* *F. verticillioides*		189
Peumus boldus	Essential oil	*A. flavus*	AFB_1	145
Pimpinella anisum	Essential oil	*A. flavus*	AFB_1	145
Pimpinella anisum	Essential oil	*A. parasiticus* *A. flavus* *A. alutaceus* *F. verticillioides*	Aflatoxins Ochratoxin A fumonisin	190
Polyminia sonchifolia	Hexanic and ethyl acetate extracts	*A. flavus*	AFB_1 AFB_2	134
Rosmarinus officinalis	Essential oil	*A. parasiticus*	AFB_1	132, 203

Table 2. (Continued)

PLANTS	EXTRACT	MOLDS	TOXIN	REFERENCE NUMBER
Satureja hortensis	Essential oil Methanolic extract	*A. parasiticus* NRLL 2999 *A. flavus*	AFB_1 AFG_1	148, 149
Syzygium aromaticum	Essential oil	*A. flavus* *A. westerdifkial*	AFB_1 AFB_2 Ocratoxina (A)	126, 145, 147, 160
Thymus eriocalyx	Essential oil	*A. parasiticus* *A. niger*	Aflatoxins	137, 204
Thymus vulgaris	Essential oil	*A. parasiticus* *P. verrucosum* *A. niger A. flavus* *A. terreus* *F. oxysporum* *Cladosporium herbarum* *A. fumigatus* *F. verticillioides* *A. alutaceus*	AFB_1 AFG_1 Ochratoxin A Fumunisin	138, 143, 147, 148, 190, 195
Thymus x-porlock	Essential oil	*A. parasiticus* *A. niger*	Aflatoxins	137, 204
Trachsperunun capticem	Essential oil	*A. flavus* *A. parasiticus*	Aflatoxins	132

Several species of *Thymus* have been reported to possess antifungal properties. Rasooli and Owlia (2005) [137]reported that the fungicidal activity of essential oils from *Thymus eriocaly* and *Thymus x-porlock* on *A. parasiticus*, demonstrated lethal effects observed at 500 and 1000 ppm respectively. The aflatoxin production was inhibited at 250 ppm of both oils. Kumar et al (2008) [138] related that the essential oil of *Thymus vulgaris* L. showed highest antifungal efficacy. The *Thymus vulgaris* oil inhibited the mycelial growth of *A. flavus* at 0.7 µL/mL and exhibited a broad fungitoxic spectrum against eight different food contaminating the fungi species: *Fusarium oxysporum*, *Cladosporium herbarum*, *Curvularia lunata*, *A. terreus*, *A. niger*, *A. fumigatus*, *Alternaria alternata* and *Botryodiploidia theobromae*. The oil also showed significant anti-aflatoxigenic efficacy as it completely arrested the aflatoxin B_1 production at 0.6 µL/mL.

Mishra (1994) [139] during screening of essential oils isolated from different parts of 14 species of higher-order plants at 3000 ppm, only the essential oil of *Cymbopogon citratus* (family Poaceae) possessed absolute toxicity against *A. flavus*. The essential oils of species of *Citrus* inhibited on the growth of moulds commonly associated with food spoilage [140]. The work showed the capacity to reduce or inhibit the growth of the moulds *Penicillium chrysogenum*, *P. verrucosum*, *A. niger* and *A. flavus* by *Citrus lemon, C. reticulate, C. paradis and C. sinensis* during the nine days. The *A. niger* and *A. flavus* growth was completely inhibited at concentration 0.94% of any of the essential oils used. The *Citrus sinensis* essential oil was most effective.

Cinnamomum zeilanicum Breym and *Allium sativum* L. essential oils were also evaluated against 37 fungal isolates for a period of 12 months and showed e highest inhibition of the mycelial development of *A. flavus* was obtained with cinnamon and garlic essential oils [141]. *C. zeilanicum* essential oil also has been inhibited the fungal growth of *A. flavus, F. graminearum, F. moniliforme (synonym F. verticillioides), P. citrinum, P. viridicatum, A. terreus* [142].

The essential oil from *Ocimum gratissimum*, *Thymus vulgaris* and *Cymbopogon citratus* were the most effective and prevented conidial germination and the growth of *Fusarium moniliforme (=F. verticillioides)*, *A. flavus* and *A. fumigatus* on corn meal agar at 800, 1000 and 1200 ppm respectively [143]. Same authors continued their studied using the same essential oils and showed that they were active against the *P. expansum*, *P. verrucosum* and *A. ochraceus*, respectively [144].

Pimpinella anisum L., *Peumus boldus* Mol, *Hedeoma multiflora* Benth, *Syzygium aromaticum* L, and *Lippia turbinate* var. integrifolia essential oils were evaluated against *Aspergillus* section *Flavi* in sterile maize grain under different *Aw*. All concentrations assayed of the plants completely inhibited fungal growth rate of the four isolates studied when *Aw* conditions were 0.95 and 0.90. Whereas *P. anisum* and *S. aromaticum* showed minor inhibitory effect [145]. These authors showed that the five essential oils tested had significant inhibitory effect on lag phase, growth rate, and aflatoxin accumulation by *Aspergillus* section *Flavi* isolates in sterile maize grain at different *Aw* conditions.

Pimpinella anisum, *Peumus boldus*, *Mentha piperita*, *Origanum vulgare* and *Minthosthachys verticillata* essential oils showed antifungal effect against *Aspergillus* section *Flavi* (two isolates of *A. parasiticus* and two isolates of *A. flavus*). *P. anisum* and *P. boldus* essential oils were most inhibitory at 500 mg/kg to all growth parameters of the fungus. These essential oils inhibited the percentage of germination, germ-tube elongation rate and fungal growth. AFB_1 accumulation was completely inhibited by *P. anisum*, *P. boldus* and *O. vulgare* essential oils. *M. piperita* and *M. verticillata* essential oils inhibited AFB_1 production by 85–90% in all concentrations assayed [146]

Thymus vulgaris, *Satureja hortensis* and *Syzygium aromaticum* essential oils were evaluated in culture medium and tomato paste. *A. flavus* were inoculated in Sabouraud Dextrose Broth and tomato paste. Results showed that all essential oils could inhibit the growth of *A. flavus*. Taste panel evaluations were carried out in a tomato ketchup base, and the percent of inhibition of each essential oil in tomato paste was lower than culture medium [147].

Others studies related to the essential oil and extract of *Satureja hortensis* against *A parasiticus* and *A. flavus in vitro* on Petri plates and liquid culture. *Satureja hortensis* L. was found as a potent inhibitor of aflatoxins B_1 (AFB_1) and G_1 (AFG_1) production by *A. parasiticus* NRRL 2999. Fungal growth was also inhibited in a dose-dependent manner [148]. Dikbas (2008) [149] also related the results of *in vitro* assays showed that the essential oil of *S. hortensis* had a strong fungicidal effect against *A. flavus* on Petri plate. The mean inhibition zone and the minimal inhibition concentration value of the oil were recorded as 61 mm and at 6.25 μL/mL.

Sygygium aromaticum, *Cymbopogon citratus*, *Cymbopogon martini*, *Cinnamomum zeylanicum* and *Origanum vulgare* essential oils were inhibitor on ZEA production by *F. graminearum*. All factors (essential oils, temperature, *Aw*, treatment timing) as well as their interactions had a significant impact on ZEA production. The *S. aromaticum*, *C. citratus* and *C. martinii* oils prevented ZEA production, while cinnamon essential oil had no significant effect, and *O. vulgare* essential oil stimulated the production [150].

The fungal growth and aflatoxin biosynthesis on *A. parasiticus* were suppressed *by Rosmarinus officinalis* and *Trachyspermum copticum* essential oils. At 5 μL of the oils per disc per Petri plate, *R. officinalis* showed inhibitory effect until 3 days of incubation while *T. copticum* L. completely inhibited the growth until 7th day of incubation. At 20 μL of the oils/disc per Petri plate, *R. officinalis* and *T. copticum* L. oils showed stronger inhibitory effects lasting for a longer period of time [132].

Ageratum conyzoides essential oil at 30 μg/mL, 20 μg/mL, 10 μg/mL, 5.0 μg/mL, 1.0 μg/mL, 0.40 μg/mL, 0.20 μg/mL, 0.10 μg/mL, 0.04μg/mL, 0.02μg/mL, 0.01 μg/mL concentrations inhibited the fungal growth on average 55%. The aflatoxin production was completely inhibited by the first eight concentrations. The study of action mechanism were realized by comparison by transmission electron microscopy (TEM) of the fungal cells, untreated and treated with essential oil, showed that the essential oil of *A. conyzoides* in different concentrations crosses not only the cell wall, but also the plasma membrane, interacting with the membranous structures of cytoplasmic organelles. Change was observed in the membranes of organelles, especially mitochondria, in which deletion and destruction of electron dense granules were noted [151].

Plants secondary metabolites with complex molecular structures and some of them have antimicrobial properties. Some works have showed that secondary metabolites include alkaloids, flavonoids, isoflavonoids, biflavonoids, anthocianidins, β-glucans, tannins, coumarins, glycosides, terpenes and phenolic compounds are biologically active against mycotoxigenic fungal (Table 3). In this panorama, spices have emerged as effective compounds to provide microbiological safety of foods. Spices are rich in essential oils, which are composed of many compounds (eugenol, citral, pinene, thymol, cinnamic acid, carvacrol) characterized by a prominent antimicrobial activity.

Luo at al (2004) [152] reported to study of citral from *Litsea cubeba* oil has been found to have a strong influence on fungi, especially *A. flavus*. Multiplex microanalysis and quasi-elastic light scattering techniques were applied to study the effects of citral on *A. flavus* spores from the levels of membrane, organelle and intracellular macromolecule. It was found that citral injured the wall and the membrane of *A. flavus* spore, resulting in decrease of its elasticity. After entering the cell, citral, not only influenced the genetic expression of mitochondrion reduplication and its morphology, but also changed the aggregation of protein-like macromolecules.

The 1,8-cineole, were evaluated for their effectiveness against the storage fungi *A. flavus* and *A. parasiticus* and showed partial inhibition only at the highest level. Moreover the essential oil from *Eucalyptus globulus* completely fungal growth inhibition of both species. Volatile exposure showed total inhibition at the lower level tested of 500 μL [153]. Probably the difference in activity is due to the synergism of the components of essential oil.

Table 3: Natural substances with inhibition activity on fungal growth and mycotoxin production.

SUBSTANCES	MOULD	TOXIN	REFERENCE NUMBER
β-glucans	*A. flavus*	AFB_1	179
1,8-cineale	*A. flavus* *A. parasiticus*	AFB_1	153
3', 4', 5- trihydroxy-7-methoxy flavanone	*A. flavus*	AFB_1	136
3', 5, 7 trihydroxy-3, 4'-dimethoxyflavone	*A. flavus*	AFB_1	136
5'-preny lisoflavanones	*A. nomius* *A. flavus*		180
6,6"-bigenkwanin, amenthoflavone	*A. flavus*	AFB_1, AFB_2	170
7,7"-dimethoxyagastisflavone	*A. flavus*	AFB_1, AFB_2	170
Anthocyanidins	*A. flavus*	AFB_1	168
Caffeic acid	*A. flavus*	AFB_1 AFG_2	181
Carvacrol	*A. parasiticus*	AFB_1, AFG_2	182
Chitosanases	*A. flavus*		166,167
Cinnamic aldehyde	*A. parasiticus*		156
Citral	*A. flavus*		152
Enhydrin and Uvedalin	*A. flavus*	AFB_1	133
Eugenol	*A. flavus* *Aspergillus* spp. *Penicillium* spp. *Fusarium* spp.	Aflatoxins, Fumonisin B_1	183, 184
Ferulic acid	*Fusarium* spp. *A. flavus* *A. parasiticus*	Tricothecene B	185, 186
limonene	*F. verticillioides*	Fumonisin B_1	165
Menthane	*F. verticillioides*	Fumonisin B_1	165
Mentol	*F. verticillioides*	Fumonisin B_1	165
o-Coumaric acid	*A. flavus*	AFB_1, AFB_2	181
Thymol	*F. verticillioides* *A. parasiticus*	Fumonisin B_1, AFB_1 AFG_1	165, 148
Trans-cinnamic acid	*A. flavus* *A. parasiticus*		186

According to Farag et al. (1989) [154] and Nychas [155] the presence of phenolic OH groups able to form hydrogen bonds with the active sites of target enzymes was thought to increase antimicrobial activity. The effects of *Syzygium aromaticum* essential oil and its principal components eugenol, on growth and mycotoxin production by some toxigenic fungal genera like *Aspergillus* spp., *Penicillium* spp. and *Fusarium* spp. had been reported

[156, 157, 158, 159, 160, 161, 162, 163]. Jayashree and Subranyam (1999) [164] reported that eugenol inhibited aflatoxin production without any significant effect on growth of the organism up to a concentration of 0.75 mMol. These authors showed that the anti-aflatoxigenic actions of eugenol may be related to inhibition of the ternary steps of aflatoxin biosynthesis involving lipid peroxidation and oxygenation. Cinnamic aldehyde and eugenol inhibited mould growth and subsequent toxin production, cinnamic aldehyde at 150 ppm and eugenol at 125 ppm [156].

Dambolena (2008) [165] relates that limonene, menthol, menthone and thymol were evaluated at final concentrations of 25, 50, 75, 150, 200, 250, 500 and 1000 μL/L of culture medium. Limonene and thymol showed the highest inhibitory effects on *F. verticillioides* development. Thus, the effects of monoterpenes on fumonisin B_1 (FB_1) biosynthesis were evaluated by using corn grain (*Zea mays*) as substrate. The monoterpenes were inserted on maize 1 day before inoculation with a conidial suspension of *F. verticillioides* to give final concentrations of 75 ppm. At this concentration, thymol was the most active inhibitor on FB_1 biosynthesis.

Polysaccharides and polypeptides to have been related as natural antifungal, some of which have been demonstrated important anti-*A. flavus* properties [166,167]

Norton (1999) [168] related that aflatoxin B_1 production was inhibited by all anthocyanidins tested, and 3-hydroxy compounds were more active than 3-deoxy forms. Monoglycosides of cyanidin were 40% less inhibitory than the aglycon, whereas a monoglucoside and a diglucoside of pelargonidin were 80 and 5%, respectively, as active as the aglycon. Of eight flavonoids tested, only kaempferol was moderately active, whereas luteolin and catechin were weakly inhibitory. Binary combinations of delphinidin and three other aflatoxin inhibitors acted independently of each other. Results with an aflatoxin pathway mutant indicated that anthocyanidin inhibition. The author reported that highest AFB_1 inhibition was obtained with 3-OH compounds, which were three times more active than the related 3-deoxy compounds. Mallozzi et al (1996) [169] also reported the inhibitor effect of flavonoids in *A. flavus*.

The biflavonoids 6,6"-bigenkwanin, amenthoflavone, 7,7"-dimethoxyagastisflavone and tetradimethoxybigenkwanin isolated from *Ouratea* species were tested for inhibitory activity on *A. flavus* cultures. Suspensions of *A. flavus* spores were inoculated into 50 mL of YES medium at different biflavonoid concentrations: 5 and 10 μg/mL for 6,6"-bigenkwanin, amenthoflavone and 7,7"-dimethoxyagastisflavone, and 5, 10, 15 and 20 μg/mL for tetradimethoxybigenkwanin. The four biflavonoids showed inhibitory activity on aflatoxin B_1 and B_2 production [170]. Phenolic compounds such as acetocyringone, syringaldehyde and sinapinic acid showed inhibitory activity on AFB_1 biosynthesis by *A. flavus* in PDA. These compounds not only inhibited AFB_1 biosynthesis, but also reduced norsolinic acid production. It is clear that phenolic compounds inhibited one or more early rather than late steps in the AFB biosynthesis pathway [171].

CONCLUSION

On despite of the methods related in this chapter, the best way of mycotoxin control is the prevention using the HACCP (Hazard Analysis and Critical Control Points) studies in all segments of food and feed channel. Good Agricultural Practices (GAP), including Good Hygienic Practices (GHP) to make sure that the harvested grains will not present a hazard to the animal and consumer health and also minimize the risk to exposure to any kind of mycotoxins.

The reduction of the mycotoxins levels to an acceptable limit for consumption must be concerned at differing stages in the supply chains; however, a complete avoidance of risks posed by mycotoxins is impractical. Ideally, the methods of the mycotoxins degradation should not only reduce the concentration of mycotoxins to safe levels but also prevent production of toxic degradation products and any reduction of the nutritional value of the treated commodity. The choice of the any method to mycotoxins control must considered the cost, facility of acquisition and implementation.

Many physical, biological and chemical procedures that prevent the formation of mycotoxins or fungal contamination represent a good way to avoid the exposure to animal feed and foods for human consumption. Physical control is difficult to use in large scale, despite the good results in the experimental conditions. Biological methods using microorganism have been a good results in mycotoxins degradation. The use of natural compounds, like essential oils, with lower toxicity of the synthetic products can be a good alternative to the mycotoxins control. Nowadays no biological system and natural compounds exist to be used in the full commercial sphere.

REFERENCES

[1] Bata, A; Lasztity, R. Detoxification of mycotoxin contamination food and feed by microorganisms. *Trends in Food Science & Techonology*, 1999, 10, 223-228.

[2] Food and Agriculture Organization (FAO) of the United Nations and International Atomic Energy Agency (IAEA) (2001). Manual on the Application of HACCP System. In: Mycotoxin Prevention and Control. pp. 7–13. *Food and Nutrition Paper.* Food and Nutrition Division, Rome, Italy.

[3] Paster, N; Pushinsky, A; Menasherov, M. Inhibitory effect of *Aspergillus niger* on the growth of *Aspergillus ochraceus* and *Aspergillus flavus*, on aflatoxin formation. *Journal of The Science of Food and Agriculture*, 1992, 58, 589–591.

[4] Samarajeewa, U; Sen,AC; Cohen,MD; Wei, CL. Detoxification of aflatoxins in food and feed by physical and chemical methods. *Journal of Food Protection*, 1990, 53,489–501.

[5] Bruyn, IN. The application of high dose food irradiation in South Africa. *Radiation Physics and Chemistry*, 2000, 57,223– 225.

[6] Lara, J; Fernandez,PS; Periago,PM; Palop, A. Irradiation of spores of *Bacillus cereus* and *Bacillus subtillus* with electron beams. *Innovative Food Science* and *Emerging Technologies*, 2002, 3, 379–384.

[7] Shapira, R; Paster, N. Control of mycotoxins in storage and techniques for their decontamination. In: Magan, N; Olsen M, editors. *Mycotoxins in Food. Detection and Control.* CRC Press, Woodhead Publishin Limited, England, 2004, 190–223.

[8] CAST Report, 2003. Mycotoxins: risks in plant, animal, and human systems. In: Richard, JL; Payne, GA, editors. *Council for Agricultural Science and Technology Task Force Report*. Ames, Iowa, USA, 2003, 139.

[9] Binder, EM. Managing the risk of mycotoxins in modern feed production. *Animal Feed Science and Technology,* 2007,133,149–166.

[10] Srivastava, PK. Prevention and control of mycotoxins in foodgrains in India. 2009-08-06. Avaiable from: http://www.fao.org/docrep/X5036E/x5036E17.htm.

[11] Samarajeewa, U. In situ degradation of mycotoxins by physical methods. In: Smith, JE. And Henderson, RS,editors. Mycotoxins and animal foods. CRC Press, 1991, 785-794.

[12] Magan, N; Aldred, D. Post-harvest control strategies: Minimizing mycotoxins in the food chain. *International Journal of Food Microbiology*, 2007, 119,131–139.

[13] Seitz, LM; Sauer, DB; Mohr, HE. Storage of high moisture corn: fungal growth and dry matter loss. *Cereal Chemistry,* 1982,59, 100–105.

[14] West DI; Bullerman, LB. Physical and Chemical separation of mycotoxins from agricultural product.In: Smith, JE. and Henderson, RS, editors. *Mycotoxins and animal foods*. CRC Press, 1991, 778-781.

[15] Herzallah, S; Alshawabkeh, K; AL Fataftah, A. Aflatoxin Decontamination of Artificially Contaminated Feeds by Sunlight, γ-Radiation, and Microwave Heating. *The Journal of Applied Poultry Research,* 2008, 17,515–521.

[16] Tsubouchi,H; Yamamoto, K; Hisada, K; Sakabe, Y; Udagawa, S. Effect of roasting on ochratoxin A level in green coffee beans inoculated with Aspergillus ochraceus. *Mycopathologia*, 1987, 97(2),111-113.

[17] Conway HF; Anderson, RA; Bagley, EB. Detoxification of aflatoxin contaminated corn by roasting. *Cereal Chemistry*, 1978, 55,115.

[18] Young, JC; Chen, ZJ; Marquardt, RR. Reduction in alkaloid content of ergot sclerotia by chemical and physical treatment. *Journal of Agricultural and Food Chemistry*, 1983,31,413.

[19] Samapundo, S; Devlieghere, F.; De Meulenaer, B.; Geeraerd, AH.; Van Impe, JF; Debevere, JM. Predictive modelling of the individual and combined effect of water activity and temperature on the radial growth of *Fusarium verticillioides* and *F. proliferatum* on corn. *International Journal of Food Microbiology*, 2005, 105(1),35-52.

[20] Frisvad, JC; Samson, RA. Filamentous fungi in food and feeds: ecology, spoilage and mycotoxin contamination. In: Arora, DK; Mukerji, KG; Marth, EH, editors. *Handbook of Applied Mycology. Foods and Feed*. Marcel Dekker, New York, 1991, 3, 61–68.

[21] Lindblad,M; Johnsson, P; Jonsson, N; Lindqvist, R; Olsen, M. 2004. Predicting noncompliant levels of ochratoxin A in cereal grain from *Penicillium verrusocum* counts. *Journal of Applied Microbiology*, 2004, 97, 609–616.

[22] Cairns-Fuller, V; Aldred, D; Magan, N. Water, temperature and gas composition interactions affect growth and ochratoxin A production by isolates of *Penicillium verrucosum* on wheat grain. *Journal of Applied Microbiology*, 2005, 99, 1215–1221.

[23] Giorni, P; Battilani, P; Pietri, A; Magan, N. Effect of a_w and CO_2 level on *Aspergillus flavus* growth and aflatoxin production in high moisture maize post-harvest. *International Journal of Food Microbiology*, 2008, 22 (1-2), 109-113.

[24] Paster, N; Bullerman, LB. Mould spoilage and mycotoxin formation in grains as controlled by physical means. *International Journal of Food Microbiology,* 1988,7, 257-265.

[25] Cairns-Fuller, V. Dynamics and control of ochratoxigenic strains of *Penicillium verrucosum* and *Aspergillus ochraceus* in the stored grain ecosystem. PhD Thesis, Cranfield University, Silsoe, U.K, 2004.

[26] Paster, N; Lisker, N; Chet, I. Ochratoxin production by *Aspergillus flavus* Wilhem grown under controlled atmospheres. *Applied and Environmental Microbiology,* 1983,45, 1136–1139.

[27] Paster, N; Blumenthal-Yonassi, J; Barkai-Golan, R; Menasherov, M. Production of zearalenone in vitro and in corn grains stored under modified atmospheres. *International Journal of Food Microbiology,* 1991,12 (2-3), 157-165.

[28] Wilson, DM; Jay, E. Effect of controlled atmosphere storage on aflatoxin production in high moisture peanuts (groundnuts). *Journal of Stored Products Research*, 1976,12(2), 97-100.

[29] Babich, H; Stotzky, G. Gaseous and heavy metal pollutants. In: Burns, RG; Slater, JH, editors. *Experimental Microbial Ecology*. Blackwell, Oxford, 1980, 631.

[30] Raghunathan, MN; Muthu, M; Majumder, SK. Control of internal fungi in sorghum by fumigation. *Journal of Stored Product Research,* 1969,5, 389–392.

[31] Eckhoff, SR; Van Cauwenberger, JE; Bothast, RJ; Nofsinger, GW; Bagley, EB. Sulphur dioxide-supplemented ambient air drying of high moisture corn. Transaction and American Society of Agricultural Engineering, 1979, 23, 1028.

[32] Eckhoff, SR; Tuite, JF; Foster, GH; Kirleis, AW; Okos, MR. Microbial growth inhibition by SO_2 plus NH_3 treatments during slow drying of corn. *Cereal Chemistry,* 1983,60, 185–188.

[33] Majumber, SK; Raghunathan, AN; Rangaswamy, JR.; Majumber, S.K., Raghunathan, A.N., Rangaswamy, J.R. Control of mycoflora on moist grain. In: Multon, JL; Guilbot, A, editors. Preservation of wet harvested grains. *Institut National de la Recherche Agronomique*, Paris, France, 1973, 343–352.

[34] Tweedie, JW; Segal, LW, Tweedie, JW; Segal, LW. Specificity of transport processes for sulphur, selenium and molybdenum anions by filamentous fungi. *Biochemistry et Biophysica Acta,* 1970, 196, 95–106.

[35] Fleurat-Lessard, F. Qualitative reasoning and integrated management of the quality of stored grain: a promising new approach. *Journal of Stored Product Research*, 2002, 38, 191–218.

[36] Magan, N; Sanchis, V; Aldred, D. Role of spoilage fungi in seed deterioration. In: Aurora, DK, editor. *Fungal Biotechnology in Agricultural, Food and Environmental Applications*. Marcell Dekker, New York, 2004, 311–323.

[37] Aidoo, KE. Posthaverst storage and preservation of tropical crops. In: Smith, JE. & Henderson, RS, editors. *Mycotoxins and animal foods*. CRC Press, 1991, 747-761.

[38] Feuell, AJ. Aflatoxins in groundnuts. IV. Problems in detoxification. *Tropical Science,* 1966, 8, 61.

[39] Sharma, A.; Chen, CR; Lan, NV. Solar-energy drying systems: A review. *Renewable and Sustainable Energy Reviews*, 2009, 13 (6-7), 1185-1210.

[40] Goyal, RK. Prevention and control of mycotoxins in foodgrains in India. *Mycotoxin prevention and control in foodgrains,* 2009-08-06. Available from: http://www.fao.org/docrep/X5036E/x5036E00.HTM

[41] Samarajeewa, U; Jayatilaka, CLV; Ranjithan, A; Gamage, TV; Arseculeratne, SN.A pilot plant for detoxification of aflatoxin B_1-contaminated coconut oil by solar irradiation. *World Journal of Microbiology* and *Biotechnology*,1985, 4, 333-343.

[42] Aziz, NH; Abd El-Aal, SS. Occurrence of aflatoxin and aflatoxigenic molds in coffee beans and decontamination by gamma-irradiation. *Journal of the Egyptian Veterinary Medical Association*, 1990, 49, 951-962.

[43] Aziz, NH; Attia, ESA; Farag, SA. Effect of gamma irradiation on the natural occurrence of *Fusarium* mycotoxins in wheat, flour, and bread. *Nahrung,* 1997, 41,34–37.

[44] Diehl, J.F. *Safety of Irradiated Foods* (2. ed. revised and expanded). New York, N.Y.: Marcel Dekker Inc., 1995, 91-115.

[45] Mahrous, SR; Aziz, NH; Shahin, AA. Influence of gamma irradiation on the occurrence of pathogenic microorganisms and nutritive value of some cereal grains. *Isotope and Radiation Research*, 2003, 35, 551–586.

[46] Refai, MK; Aziz, NH; El-Far, FM; Hassan, AA. Detection of ochratoxin produced by *A. ochraceus* in feedstuffs and its control by gamma irradiation. *Applied Radiation and Isotopes,* 1996, 7, 617–621.

[47] Aziz, NH; Moussa, AA. Influence of gamma-radiation on mycotoxin producing moulds and mycotoxins in fruits. *Food Control*, 2002, 13, 281-288.

[48] Pasteur, N; Bullerman, LB. Mould spoilage and mycotoxin formation in grains as controlled by physical means. *International Journal of Food Microbiology,* 1988, 7, 257-265.

[49] Yun, H; Lim, S; Jo, C.; Chung, J; Kim, S.; Kwon, et al. Effects of organic acids an ethanol on the radio-degradation of patulin in an aqueous model system. *Radiation Physics and Chemistry,* 2008,77(6), 830-834.

[50] Mitchell, GE. Influence of irradiation of food on aflatoxin production. *Food Technology in Australia*, 1988, 40, 324-326.

[51] Van Dyck, PJ; Tobback, P; Feys, M; Van De Voorde, H. Sensitivity of Aflatoxin B_1 to ionizing radiation. *Applied and Environmental Microbiology,* 1982, 43,1317-1319.

[52] Aquino, S; Ferreira, F; Ribeiro, DHB; Corrêa, B; Greiner, R; Villavicencio, ALCH Evaluation of viability of *Aspergillus flavus* and aflatoxins degradation in irradiated samples of maize. *Brazilian Journal of Microbiology,* 2005, 36, 352-356.

[53] Aziz, NH; El-Far, FM; Shahin, AAM; Roushy, SM. Control of *Fusarium* moulds and fumonisin B_1 in seeds by gamma-irradiation. *Food Control*, 2007, 18 (11), 1337-1342.

[54] Ferreira-Castro, FL; Aquino, S; Greiner, R; Ribeiro, DHB; Reis, TA; Correa, B. Effects of gamma radiation on maize samples contaminated with *Fusarium verticillioides*. *Applied Radiation and Isotopes,* 2007, 65, 927–933.

[55] Sharma, A; Behere, AG; Padwal Desai, SR; Nadkarni, GB. Influence of inoculum size of *Aspergillus parasiticus* spores on aflatoxin production. *Applied and Environmental Microbiology,* 1980, 40, 989–993.

[56] Odamtten, GT; Appiah, V; Langerak, DI. Influence of inoculum size of *Aspergillus flavus* link on the production of aflatoxin B_1 in maize medium before and after exposure

to combination treatment of heat and gamma radiation. *International Journal of Food Microbiology*, 1987, 4,119–127.

[57] Karunaratne, A; Bullerman, LB. Interactive effects of spore load and temperature on aflatoxin production. *Journal of Food Protection*, 1990, 53, 227–229.

[58] Schatzmayr, G; Zehner, F; Täubel, M; Schatzmayr, D; Klimitsch, A; Loibner, AP; Binder EM. Microbiologicals for deactivating mycotoxins. *Molecular Nutrition & Food Reserch*, 2006, 50, 543-551

[59] Holmes, RA; Boston, RS; Payne, GA. Diverse inhibitors of aflatoxin biosynthesis. *Applied Microbiology and Biotechnology*, 2008, 78, 559-572.

[60] Wu, Q; Jezkova, A; Yuan, Z; Pavlikova, L; Dohnal, V; Kuca, K. Biological degradation of aflatoxins. *Drug Metabolism Review*, 2009, 41, 1-7.

[61] Yin, YN; Yan, LY; Jiang, JH; Ma, ZH. Biological control of aflatoxin contamination of crops. Journal of Zhejiang University Science B, 2008, 9, 787-792.

[62] El-Nezami, H; Kankaanpää, P; Salminen, S; Ahokas, J. Ability of dairy strains of lactic acid bacteria to bind a common food carcinogen, afltoxin B_1. *Food Chemistry and. Toxicology*, 1998, 36, 321–326.

[63] El-Nezami, H; Mykkanen, H; Kankaanpaa, P; Salminen, S; Ahokas, J. Ability of *Lactobacillus* and *Propionibacterium* strains to remove aflatoxin B1 from the chicken duodenum. *Journal of Food Protection*, 2000, 63, 549-552.

[64] Pierides, M; El-Nezami, H; Peltonen, K; Salminen, S; Ahokas, J. Ability of dairy strains of lactic acid bacteria to bind aflatoxin M1 in a food model. *Journal of Food Protection*, 2000, 63, 645-650.

[65] Peltonen, K; El-Nezami, H; Salminen, S; Ahokas, J. Binding of aflatoxin B1 by probiotic bacteria. *Journal of the Science of Food and Agriculture*, 2000, 80, 1942-1945.

[66] Haskard, CA; El-Nezami, HS; Kankaanpää, PE; Salminen, S; Ahokas, JT. Surface binding of aflatoxin B(1) by lactic acid bacteria. *Applied and Environmental Microbiology*, 2001, 67, 3086-3091.

[67] Peltonen, K; El-Nezami, H; Haskard, C; Ahokas, J; Salminen, S. Aflatoxin B1 binding by dairy strains of lactic acid bacteria and B*ifidobacteria*. *Journal of Dairy Science*, 2001, 84, 2152-2156.

[68] Taylor, WJ; Draughon, FA. *Nannocystis exedens*: a potential biocompetitive agent against *Aspergillus flavus* and *Aspergillus parasiticus*. Journal of Food Protection, 2001, 64, 1030-1034.

[69] Hormisch, D; Brost, I; Kohring, GW; Gifhorn, F; Kroppenstedt, RM; Stackebrandt, E; Färber, P; Holzapfel, WH. *Mycobacterium fluoranthenivorans* sp. nov., a fluoranthcnc and aflatoxin B1 degrading bacterium from contaminated soil of a former coal gas plant. *Systematic and Applied Microbiology*, 2004, 27, 653–660.

[70] Shapira, R. Detection and control. In: Magan, N; Olsen, M (editor). *Mycotoxins in Food*. Bocca Raton, Florida, USA: CRC Press. 2004, 207.

[71] Gratz, S; Mykkänen, H; El-Nezami, H. Aflatoxin B1 binding by a mixture of *Lactobacillus* and *Propionibacterium*: in vitro versus ex vivo. *Journal of Food Protection*, 2005, 68, 2470-2474.

[72] Alberts, JF; Engelbrecht, Y; Steyn, PS; Holzapfel, WH; Van Zyl, WH. Biological degradation of aflatoxin B1 by *Rhodococcus erythropolis* cultures. *International Journal of Food Microbiology*, 2006, 109, 121-126.

[73] Khanafari, A; Soudi, H; Miraboulfathi, M; Osboo, RK. An *in vitro* investigation of aflatoxin B1 biological control by *Lactobacillus plantarum. Pakistan Journal of Biological Science*, 2007, 10, 2553-2556.

[74] Palumbo, JD; O'Keeffe, TL; Abbas, HK. Isolation of maize soil and rhizosphere bacteria with antagonistic activity against *Aspergillus flavus* and *Fusarium verticillioides. Journal of Food Protection*, 2007, 70, 1615-1621.

[75] Guan, S; Ji, C; Zhou, T; Li, J; Ma, Q; Niu, T. Aflatoxin B(1) degradation by *Stenotrophomonas maltophilia* and other microbes selected using coumarin medium. *International Journal of Molecular Sciences*, 2008, 9, 1489-1503.

[76] Tejada-Castañeda, ZI; Avila-Gonzalez, E; Casaubon-Huguenin, MT; Cervantes-Olivares RA; Vásquez-Peláez, C; Hernández-Baumgarten, EM; Moreno-Martínez, E. Biodetoxification of aflatoxin-contaminated chick feed. *Poultry Science*, 2008, 87, 1569-1576.

[77] Fazeli, MR; Hajimohammadali, M; Moshkani, A; Samadi, N; Jamalifar, H; Khoshayand, MR; Vaghari, E; Pouragahi, S. Aflatoxin B1 binding capacity of autochthonous strains of lactic acid bacteria. *Journal of Food Protection*, 2009, 72, 189-192.

[78] Kabak, B; Var, I. Factors affecting the removal of aflatoxin M1 from food model by *Lactobacillus* and *Bifidobacterium* strains. *Journal of Environmental Science and Health B*, 2008, 43, 617-624.

[79] La Penna, M; Nesci, A; Etcheverry, M. In vitro studies on the potential for biological control on *Aspergillus* section Flavi by *Kluyveromyces* spp. *Letters Applied in Microbiology*, 2004, 38, 257-264.

[80] Shantha, T. Fungal degradation of aflatoxin B1. *Natural Toxins*, 1999, 7, 175-178.

[81] Chang, PK; Hua, SS. Nonaflatoxigenic *Aspergillus flavus* TX9-8 competitively prevents aflatoxin accumulation by *A. flavus* isolates of large and small sclerotial morphotypes. *International Journal of Food Microbiology*, 2007, 114, 275-279.

[82] Cotty, PJ. Influence of field application of an atoxigenic strain of *Aspergillus flavus* on the populations of *A. flavus* infecting cotton bolls and on the aflatoxin content of cottonseed. *Phytopathology*, 1994, 84, 1270-1277.

[83] Pitt, JI; Hocking, AD. Mycotoxins in Australia: biocontrol of aflatoxin in peanuts. *Mycopathologia*, 2006, 162, 233-243.

[84] Dorner, JW; Horn, BW. Separate and combined applications of nontoxigenic Aspergillus flavus and A. parasiticus for biocontrol of aflatoxin in peanuts. *Mycopathologia*, 2007, 163, 215-223.

[85] Atehnkeng, J; Ojiambo, PS; Ikotun, T; Sikora, RA; Cotty, PJ; Bandyopadhyay, R. Evaluation of atoxigenic isolates of *Aspergillus flavus* as potential biocontrol agents for aflatoxin in maize. *Food Additives and Contaminants. Part A, Chemistry, Analysis, Control, Exposure & Risk Assessment*, 2008, 25, 1264-1271.

[86] Dorner JW. Biological control of aflatoxin contamination in corn using a nontoxigenic strain of *Aspergillus flavus. Journal of Food Protection*, 2009, 72, 801-804.

[87] Smith, JE; Moss, MO. *Mycotoxins. Formation, Analysis and Significance*. New York, Chichester: John Wiley and Sons Ltd; 1985.

[88] Petchkongkaew, A; Taillandier, P; Gasaluck, P; Lebrihi, A. Isolation of *Bacillus* spp. from Thai fermented soybean (Thua-nao): screening for aflatoxin B1 and ochratoxin A detoxification. *Journal Applied and Microbiology*, 2008, 104, 1495-1502.

[89] Piotrowska, M; Zakowska, Z. The elimination of ochratoxin A by lactic acid bacteria strains. *Polish Journal of Microbiology*, 2005, 54, 279-286.

[90] Fuchs, S; Sontag, G; Stidl, R; Ehrlich, V; Kundi, M; Knasmüller, S. Detoxification of patulin and ochratoxin A, two abundant mycotoxins, by lactic acid bacteria. *Food Chemistry and Toxicology*, 2008, 46, 1398-13407.

[91] De Felice, DV; Solfrizzo, M; De Curtis, F; Lima, G; Visconti, A; Castoria, R. Strains of *Aureobasidium pullulans* can lower ochratoxin A contamination in wine grapes Phytopathology, 2008, 98, 1261-1270.

[92] Abrunhosa, L; Serra, R; Venâncio, A. Biodegradation of ochratoxin A by fungi isolated from grapes. *Journal Agriculture and Food Chemistry*, 2002, 50, 7493-7496.

[93] Varga, J; Rigó, K; Téren, J. Degradation of ochratoxin A by *Aspergillus* species. *International Journal of Food Microbiology*, 2000, 59, 1-7.

[94] Varga, J; Péteri, Z; Tábori, K; Téren, J; Vágvölgyi, C. Degradation of ochratoxin A and other mycotoxins by *Rhizopus* isolates. *International Journal Food of Microbiology*, 2005, 99, 321-328.

[95] Bleve, G; Grieco, F; Cozzi, G; Logrieco, A; Visconti, A. Isolation of epiphytic yeasts with potential for biocontrol of *Aspergillus carbonarius* and *A. niger* on grape. *International Journal of Food Microbiology*, 2006, 108, 204-209.

[96] Karlovsky, P. Biological detoxification of fungal toxins and its use in plant breeding, feed and food production. *Natural Toxins*, 1999, 7, 1– 23.

[97] Castoria, R; Morena, V; Caputo, L; Panfili, G; De Curtis, F; De Cicco, V. Effect of the biocontrol yeast *Rhodotorula glutinis* strain ls11 on patulin accumulation in stored apples. *Phytopathology*, 2005, 95, 1271-1278.

[98] Morales, H; Sanchis, V; Usall, J; Ramos, AJ; Marín, S. Effect of biocontrol agents *Candida sake* and *Pantoea agglomerans* on *Penicillium expansum* growth and patulin accumulation in apples. *International Journal of Food Microbiology*, 2008, 122, 61-67.

[99] Snook ME; Mitchell T; Hinton DM; Bacon CW. isolation and characterization of leu(7)-surfactin from the endophytic bacterium *bacillus mojavensis* rrc 101, a biocontrol agent for *fusarium verticillioides*. *Journal Agriculture and Food Chemistry*, 2009, 57, 4287-4292.

[100] Bacon, CW; Yates, IE; Hinton, DM; Meredith, F. Biological control of *Fusarium moniliforme* in maize. *Environmental Health Perspectives*, 2001, 109, Suppl 2, 325-332.

[101] Cavaglieri, L; Orlando, J; Rodríguez, MI; Chulze, S; Etcheverry, M. Biocontrol of *Bacillus subtilis* against *Fusarium verticillioides* in vitro and at the maize root level. *Research in Microbiology*, 2005, 156, 748-754.

[102] Gwiazdowska, D; Czaczyk, K; Filipiak, M; Gwiazdowski, R. Effects of Propionibacterium on the growth and mycotoxin production by some species of *Fusarium* and *Alternaria*. *Polish Journal of Microbiology*, 2008, 57, 205-212.

[103] Cavaglieri, L; Orlando, J; Etcheverry, M. In vitro influence of bacterial mixtures on *Fusarium verticillioides* growth and fumonisin B1 production: effect of seeds treatment on maize root colonization. *Letters Applied of Microbiology*, 2005, 41, 390-396.

[104] Cavaglieri, L; Passone, A; Etcheverry, M. Screening procedures for selecting rhizobacteria with biocontrol effects upon *Fusarium verticillioides* growth and fumonisin B1 production. Research in Microbiology, 2004, 155, 747-754.

[105] Calistru, C; McLean, M; Berjak, P. In vitro studies on the potential for biological control of *Aspergillus flavus* and *Fusarium moniliforme* by *Trichoderma* species. A study of the production of extracellular metabolites by *Trichoderma* species. *Mycopathologia*, 1997, 137, 115-124.

[106] Yates, IE; Meredith, F; Smart, W; Bacon, CW; Jaworski, AJ. *Trichoderma viride* suppresses fumonisin B1 production by *Fusarium moniliforme. Journal of Food Protection*, 1999, 62, 1326-1332.

[107] Shima, J; Takase, S; Takahashi, Y; Iwai, Y; Fujimoto, H; Yamazaki, M; Ochi, K. Novel detoxification of the trichothecene mycotoxin deoxynivalenol by a soil bacterium isolated by enrichment culture. *Applied and Environmental Microbiology*, 1997, 63, 3825-3830.

[108] El-Nezami, HS; Chrevatidis, A; Auriola, S; Salminen, S; Mykkänen, H. Removal of common *Fusarium* toxins in vitro by strains of *Lactobacillus* and *Propionibacterium. Food Additives and Contaminants*, 2002, 19, 680-686.

[109] Schisler, DA; Khan, NI; Boehm, MJ. Biological control of *Fusarium* head blight of wheat and deoxynivalenol levels in grain via use of microbial antagonists. *Advances in Experimental Medicine and Biology*, 2002, 504, 53-69.

[110] Völkl, A; Vogler, B; Schollenberger, M; Karlovsky, P. Microbial detoxification of mycotoxin deoxynivalenol. *Journal of Basic Microbiology*, 2004, 44, 147-156.

[111] Beeton, S; Bull, AT. Biotransformation and detoxification of T-2 toxin by soil and freshwater bacteria. *Applied and Environmental Microbiology*, 1989, 55, 190-197.

[112] Hassan, YI; Bullerman, LB. Antifungal activity of *Lactobacillus paracasei* subsp. tolerans against *Fusarium proliferatum* and *Fusarium graminearum* in a liquid culture setting. *Journal of Food Protection*, 2008, 71, 2213-2216.

[113] Czaczyk, K; Trojanowska, K; Mueller, A. Antifungal activity of *Bacillus coagulans* against *Fusarium* spp. *Acta Microbiologuca Polonica*, 2002, 51, 275-283.

[114] Zinedine, A; Soriano, JM; Moltó, JC; Mañes, J. Review on the toxicity, occurrence, metabolism, detoxification, regulations and intake of zearalenone: an oestrogenic mycotoxin. *Food Chemistry and Toxicology*, 2007, 45, 1-18.

[115] Namazi, M; Allameh, A; Aminshahidi, M; Nohee, A; Malekzadeh, F. Inhibitory effects of ammonia solution on growth and aflatoxin production by Aspergillus parasiticus NRRL-2999. *Acta Poloniae Toxicologica*, 2002, 10, 65–72.

[116] Galvano, F; Piva, A; Ritteni, A; Galvano, G. Dietary strategies to counteract the effects of mycotoxins: a review. *Journal of Food Protection*, 2001 64, 120–131.

[117] Sharma; RP; Salunkhe, DK. Detoxification of Mycotoxins. In Sharma, RP; Salunkhe, DK (editors). *Mycotoxins and Phytoalexins*. Florida, US: CRC Press, Inc., 1991; pp. 464.

[118] Norred, WP; Voss, KA; Bacon, CW; Riley, RT. Effectiveness of ammonia treatment in detoxification of fumonisin-contaminated corn. *Food Chemistry and Toxicology*, 1991, 29, 815–819.

[119] Park, DL. Controlling aflatoxin in food and feed. *Food Technology* 1993. 47, 92–96.

[120] Young, JC; Trenholm, HL; Friend, DW; Prelusky, DB. Detoxification of deoxynivalenol with sodium bisulfite and evaluation of the effects when pure mycotoxin or contaminated corn was treated and given to pigs. *Journal Agriculture and Food Chemistry*, 1987, 35, 259–261.

[121] Kappes, E.M; Serrati, L; Drouillard, JB. A crop protection approach to Aspergillus and OTA management in Southren European vineyards. Proceedings from the International Workshop Ochratoxin A in Grapes and Wine: Prevention and Control. Marsala, Italy, pp. 24.

[122] Marín, S; Ramos, AJ; Sanchis, V. Chemical Control of Mycotoxigenic Fungi in R. Barkai-Golan RB; Pastor N; Editor. *Mycotoxins in Fruits and Vegetables* Academic Press; 2008; 279-296.

[123] Bellı, N; Marın, S; Sanchis, V; Ramos, AJ. Impact of fungicides on Aspergillus carbonarius growth and ochratoxin A production on synthetic grapelike medium and on grapes. *Food Additives and Contamination*, 2006, 23, 1021–1029.

[124] Barberis, C; Astoreca, A; Asili, R; Fernandez-Juri, G; Chulze, S; Magnoli, C; Dalcero, A. In vitro control of growth and ochratoxin A production by butylated hydroxyanisole in *Aspergillus* section *Nigri* species. *Food Contro*l, 2009, 20, 709-715.

[125] Paterson, RRM. Some fungicides and growth inhibitor/biocontrol-enhancer 2-deoxy-D-glucose increase patulin from *Penicillium expansum* strains in vitro. *Crop Protection*, 2007, 26, 543–548.

[126] Reddy, KRN; Reddy, CS; Muralidharan, K. Potential of botanicals and biocontrol agents on growth and aflatoxin production by *Aspergillus flavus* infecting rice grains. *Food Control*, 2009, 20, 173-178.

[127] Méndez-Albores, A; J. Veles-Medina, J; Urbina-Álvarez, E; Martínez-Bustos, F; Moreno-Martínez, E. Effect of citric acid on aflatoxin degradation and on functional and textural properties of extruded sorghum. *Animal Feed Science and Technology*, 2009, 150, 316-329.

[128] Wu, HS; Raza, W; Fan, JQ; Sun, YG; Bao, W; Liu, DY; Huang, OW; Mao, ZS; Shen, OR; Miao, WG. Antibiotic effect of exogenously applied salicylic acid on *in vitro* soilborne pathogen, *Fusarium oxysporum* f.sp.*niveum*. *Chemosphere*, 2008, 74, 45-50.

[129] Pandey, A. Solid state fermentation. *Biochemical Engineering Journal*, 2003, 13, 81-84.

[130] Kumar, R; Mishra, AK; Dubey, NK; Tripathi, YB. Evaluation of *Chenopodium ambrosioides* oil as a potential source of antifungal, antiaflatoxigenic and antioxidant activity. *International Journal of Food Microbiology*, 2007, 115, 159-164.

[131] McKenzie, KS; Sarr, AB; Mayura, K; Bailey, RH; Miller, DR; Rogers, TD; Norred, WP; Voss, KA; Plattner, RD; Kubena, LF; Phillips, TD. Oxidative degradation and detoxification of mycotoxins using a novel source of ozone. *Food Chemistry and Toxicology*, 1997, 35, 807–820.

[132] Rasooli I; Fakoor MH; Yadegarinia D; Gachkar L; Allameh A; Rezaei MB. Antimycotoxigenic characteristics of *Rosmarinus officinalis* and *Trachyspermum copticum* L. essential oils, *International Journal of Food Microbiology*, 2008, 122, 135-139

[133] Ghorbanian, M; Razzaghi-Abyaneh, M; Allameh, A; Shams-Ghahfarokhi, M; Qorbani, M. Study on the effect of neem (*Azadirachta indica* A. juss) leaf extract on the growth

of *Aspergillus parasiticus* and production of aflatoxin by it at different incubation times. *Mycoses*, 2007, 51, 35–39.

[134] Gonçalez, E; Felicio, JD; Pinto, MM; Rossi, MH; Medina, C; Fernandes, MJB; Simoni, IC. Inhibition of aflatoxin production by *Polymnia sonchifolia* and its *in vitro* cytotoxicity. *Arquvos do Instituto. Biologico*, 2003, 70, 159-163.

[135] Pinto, MM; Gonçalez, E; Rossi, MH; Felício, JD; Medina, CS; Fernandes, MJB; Simoni, IC. Activity of the aqueous extract from *Polymnia sonchifolia* leaves on growth and production of aflatoxin B1 by *Aspergillus flavus*, *Brazilian Journal Microbiology,* 2001, 32, 127-129.

[136] Pak, A; Gonçalez, E; Felício, JD; Pinto, MM; Rossi, MH; Simoni, IC; Lopes, MN. Inhibitory activity of compounds isolated from *Polymnia sonchifolia* on aflatoxin production by *Aspergillus flavus*. *Brazilian Journal Microbiology*, 2006, 37, 199-203.

[137] Rasooli, I; Owlia P. Chemoprevention by thyme oils of *Aspergillus parasiticus* growth and aflatoxin production. *Phytochemistry*, 2005, 66, 2851-2856.

[138] Kumar, A; Shukla, R; Singh, P; Prasad, CS; Dubey, NK. Assessment of *Thymus vulgaris* L. essential oil as a safe botanical preservative against post harvest fungal infestation of food commodities. *Innovative Food Science & Emerging Technologies*, 2008, 9, 575-580.

[139] Mishra, AK; Dubey, NK. Evaluation of some essential oils for their toxicity against fungi causing deterioration of stored food commodities. *Applied and Environmental Microbiology*, 1994, 60, 1101-1105.

[140] Viuda-Martos, M; Ruiz-Navajas, Y; Fernández-López, J; Perez-Álvarez, J. Antibacterial activity of lemon (Citrus lemon l.), mandarin (Citrus reticulata l.), grapefruit (Citrus paradisi l.) and orange (Citrus sinensis l.) essential oils. *Journal of Food Safety*, 2009, 28, 567–576.

[141] Viegas, EC; Soares, A; Carmo, MGF; Rossetto, CAV. Toxicidade de óleos essenciais de alho e casca de canela contra fungos do grupo *Aspergillus flavus*. *Horticultura Brasileira*, 2005, 23, 915-919.

[142] Singh, HNP; Prasad, MM; Sinha, KK. Efficacy of leaf extracts of some medicinal plantas against disease deve lop ment in banana. *Letters in Applied Microbiology,* 1993, 17, 269-271.

[143] Nguefack, J; Leth, V; Zollo, PHA; Mathur, SB. Evaluation of five essential oils from aromatic plants of Cameroon for controlling food spoilage and mycotoxin producing fungi. *International journal of food microbiology*, 2004, 94, 329-334.

[144] Nguefack, J; Dongmo, JBL; Dakole, CD; Leth, V; Vismer, HF; Torp, J; Guemdjom, EFN; Mbeffo, M; Tamgue, O; Fotio, D; Zollo, PHA; Nkengfack, AE. Food preservative potential of essential oils and fractions from *Cymbopogon citratus*, *Ocimum gratissimum* and *Thymus vulgaris* against mycotoxigenic fungi. *International Journal of Food Microbiology*, 2009, 131, 151-156.

[145] Bluma, RV; Etcheverry, MG. Application of essential oils in maize grain: Impact on *Aspergillus* section *Flavi* growth parameters and aflatoxin accumulation. *Food Microbiology*, 2008, 25, 324-334.

[146] Bluma, RV; Amaiden, MR; Daghero, J; Etcheverry, MG. Control of Aspergillus section Flavi growth and aflatoxin accumulation by plant essential oils. *Journal of Applied Microbiology*, 2008, 105, 203–214.

[147] Omidbeygi, M; Barzegar, M; Hamidi, Z; Naghdibadi, H. Antifungal activity of thyme, summer savory and clove essential oils against *Aspergillus flavus* in liquid medium and tomato paste. *Food Control*, 2007, 18, 1518-1523.

[148] Razzaghi-Abyaneh, M; Ghahfarokhi, S; Yoshinari, T; Rezaee, MB; Jaimand, K; Nagasawa, H; Sakuda, S. Inhibitory effects of *Satureja hortensis* L. essential oil on growth and flatoxin production by *Aspergillus parasiticus International Journal of Food Microbiology*, 2008, 123, 228-233.

[149] Dikbas, N; Kotan, K; Dadasoglu, F; Sahin, F. Control of *Aspergillus flavus* with essential oil and methanol extract of *Satureja hortensis*. *International Journal of Food Microbiology*, 2008, 124, 179–182.

[150] Marın, S; Velluti, A; Ramos, AJ; Sanchis, V. Effect of essential oils on zearalenone and deoxynivalenol production by *Fusarium graminearum* in non-sterilized maize grain. *Food Microbiology*, 2004, 21, 313–318.

[151] Nogueira, *JHC*; Gonçalez, E; Galleti, SR; Facaneli, R; Marques, MOM; Felício, JD. *Ageratum* conyzoides essential oil as aflatoxin suppressor by *Aspergillus flavus*. *International Journal of Food Microbiology*, 2009, DOI information: 10.1016/j.ijfoodmicro.2009.10.017.

[152] Luo, M; Jiang, LK; Huang, YX; Xiao, M; Li, B; Zou, GL. Effects of citral on *Aspergillus flavus* spores by quasi-elastic light scattering and multiplex microanalysis techniques. *Acta Biochimica et Biophysica Sininica*, 2004, 36, 277-83.

[153] Vilela, GR; Almeida GS; D'Arce, MABR; Moraes, MHD; Brito, JO; Silva, MFGF; Silva, SC; Piedade, SMS; Calori-Domingues, MA; Gloria, EM. Activity of essential oil and its major compound, 1,8-cineole, from *Eucalyptus globulus* Labill., against the storage fungi *Aspergillus flavus* Link and *Aspergillus parasiticus* Speare. *Journal of Stored Products Research*, 2009, 45, 108-111.

[154] Farag, RS; Daw, ZY; Hewedi, FM; El-Baroty, GSA. Antimicrobial activity of some Egyptian spice essential oils. *Journal of Food Protection*, 1989,52, 665–667.

[155] Nychas, GJE. Natural antimicrobial from plants. In: Gould, G.W. (Ed.), *New Methods of Food Preservations*, 1995, Blakie Academic and Professional, Glasgow, pp. 58–89.

[156] Bullerman, LB; Lieu, FY; Seier, SA. Inhibition of growth and aflatoxin production by cinnamon and clove oils. cinnamic aldehyde and eugenol. *Journal of Food Science*, 1977, 42, 1107 – 1109.

[157] Hitokoto, H; Morozumi, S; Wauke, T; Sakai, S; Kurata, H. Inhibitory effects of spices on growth and toxin production of toxigenic fungi. *Applied and Environmental Microbiology*, 1980, 39, 818–822.

[158] Kumar, R; Dubey, NK; Tiwari, OP; Tripathi, YB; Sinha, KK. Evaluation of some essential oils as botanical fungitoxicants for the protection of stored food commodities from fungal infestation. *Journal of the Science of Food and Agriculture*, 2007, 87, 1737–1742.

[159] Sinha, KK; Sinha, AK; Prasad, G. The effect of clove and cinnamon oils on growth of and aflatoxin production by *Aspergillus flavus*. *Letters in Applied Microbiology*, 1993, 16, 114–117.

[160] Aldred, D; Cairns-Fuller, V; Magan, N. Environmental factors affect efficacy of some essential oils and resveratrol to control growth and ochratoxin A production by *Pennicillium verrucosum* and *Aspergillus westerdijkae* on wheat grain. *Journal of Stored Products Research*, 2008, 44, 341-346.

[161] Hope, R; Jestoi, M; Magan, N. Multitarget environmental approach for control of growth and toxin production by Fusarium culmorum using essential oil and antioxidant. In: Credland, P.F., Armitage, D.M., Bell, C.H., Cogan, P.M., Highley, E. (Eds.), *Advances in Stored Product Protection. Proceedings of Eighth International Working conference on Stored Product Protection (IWCSPP),* 2002, CABI Publishing, New York, pp. 486–492.

[162] Velluti, A; Sanchis, V; Ramos, AJ; Ergido, J; Marın, S. Inhibitory effect of cinnamon, clove, lemongrass, oregano and palmarosa essential oils on growth and fumonisin B1 production by *Fusarium proliferatum* in maize grain. *International Journal of Food Microbiology*, 2003, 89, 145–154.

[163] Velluti, A; Sanchis, V; Ramos, AJ; Marın, S. Effect of essential oil of cinnamon, clove, lemongrass, oregano and palmarosa on growth and fumonisin B1 production by Fusarium verticilloides in maize grain,. *Journal of the Science of Food and Agriculture*, 2004, 89, 145–154.

[164] Jayashree, T; Subramanyam, C. Antiaflatoxigenic activity of eugenol is due to inhibition of lipid peroxidation. *Letters in Applied Microbiology*, 1999, 28, 179–183.

[165] Dambolena, JS; López, AG; Cánepa, MC; Theumer, MG; Zygadlo, JA; Rubinstein, HR. Inhibitory effect of cyclic terpenes (limonene, menthol, menthone and thymol) on *Fusarium verticillioides* MRC 826 growth and fumonisin B1 biosynthesis, *Toxicon*, 2008, 51, 37-44.

[166] Cuero, RG; Osuji, GO. *Aspergillus flavus* – induced chitosane in germinating corn and peanut seeds: *A. flavus* mechanism for growth dominance over associated fungi and concomitant aflatoxin production. *Food additives and contaminants*, 1995, 12, 479-483.

[167] Zhang, B; Wang, DF; Li, HY; Xu, Y; Zhang, L. Preparation and properties of chitosan–soybean trypsin inhibitor blend film with anti-*Aspergillus flavus* activity. *Industrial Crops and Products*, 2009, 29, 541-548.

[168] Norton, RA. Inhibition of aflatoxin B(1) biosynthesis in *Aspergillus flavus* by anthocyanidins and related flavonoids. *Journal Agriculture and Food Chemistry*, 1999, 47, 1230-1235.

[169] Mallozzi, MAB; Corrêa, B; Haraguchi, M; Brignani, FN. Effect of flavonoids on *Aspergillus flavus* growth and aflatoxin production, *Revista de Microbiologia.*, 1996, 27, 161-167.

[170] Gonçalez, E; Felício, JD; Pinto, MM. Biflavonoids inhibit the production of aflatoxin by *Aspergillus flavus*, *Brazilian Journal of Medical and Biological Research* 2001, 34, 1453- 1456.

[171] Hua, SST; Grosjean, OK; Baker, JL. Inhibition of aflatoxin biosynthesis by phenolic compounds. *Letters in Applied Microbiology,* 1999, 29, 289–291.

[172] El-Ghaouth, A; Wilson, CL; Wisniewski, M. Antifungal Activity of 2-Deoxy-D-Glucose on *Botrytis cinerea, Penicillium expansum*, and *Rhizopus stolonifer*: Ultrastructural and Cytochemical Aspects. *Phytopatology*, 1997, 87, 772-779.

[173] Razzaghi-Abyaneh, M; Shams-Ghahfarokhi, M; Kawachi, M; Eslamifar, A; Schmidt, OJ; Schmidt, A; Allameh, A; Yoshinari, T. Ultrastructural evidences of growth inhibitory effects of a novel biocide, Akacid$^{®plus}$, on an aflatoxigenic *Aspergillus parasiticus*. *Toxicon*, 2006, 48, 1075-1082.

[174] Passone, MA; Resnik, S; Etcheverry, MG. The potential of food grade antioxidants in the control of *Aspergillus* section *Flavi*, interrelated mycoflora and aflatoxin B_1 accumulation on peanut grains, *Food Control*, 2008, 19, 364-371.

[175] Medina, A.; Mateo, R; Francisco, M. Valle-Algarra, FM; Mateo, EM; Jiménez, M. Effect of carbendazim and physicochemical factors on the growth and ochratoxin A production of *Aspergillus carbonarius* isolated from grapes *International Journal of Food Microbiology*, 2007, 119, 230-235.

[176] Cabanas, R; Abarca, ML; Bragulat, MR; Cabañes, FJ. *In vitro* activity of imazalil against *Penicillium expansum*: Comparison of the CLSI M38-A broth microdilution method with traditional techniques, *International Journal of Food Microbiology*, 2009, 129, 26-29.

[177] Gardini, F; Lanciotti, R; Guerzoni, ME. Effect of *trans*-2-hexenal on the growth of *Aspergillus flavus* in relation to its concentration, temperature and water activity. *Letters in Applied Microbiology*, 2001, 33, 50-55.

[178] Neri, F; Mari, M; Menniti, AM; Brigati, S. Activity of trans-2-hexenal against Penicillium expansum in 'Conference' pears. *Journal of applied microbiology*, 2006, 100,1186-93.

[179] Zjalic, S; Reverberi, M; Ricelli, A; Granito, VM; Fanelli, C; Fabbri, AA. Trametes versicolor: A possible tool for aflatoxin control. *International Journal of Food Microbiology*, 2006, 107, 243 – 249.

[180] Quiroga, EN; Sampietro, DA; Sgariglia, MA; Soberón, JR; Vattuone, MA. Antimycotic activity of 5′-prenylisoflavanones of the plant *Geoffroea decorticans*, against *Aspergillus* species. *International Journal of Food Microbiology*, 2009, 132, 42-46.

[181] Paster, N; Juven, BJ; Harshemesh, H. Antimicrobial activity and inhibition of aflatoxin B_1 formation by olive plant tissue constituents. *Journal of Applied Microbiology*, 1988, 64, 293–297.

[182] Razzaghi-Abyaneh, M; Shams-Ghahfarokhi, M; Rezaee, MB; Jaimand, K; Alinezhad, S; Saberi, R; Yoshina, T. Chemical composition and antiaflatoxigenic activity of *Carum carvi* L., *Thymus vulgaris* and *Citrus aurantifolia* essential oils. *Food Control*, 2009, 20, 1018-1024.

[183] Guarisco, JA; Hall, JO; Coulombe Jr, RA. Mechanisms of butylated hydroxytoluene chemoprevention of aflatoxicosis - inhibition of aflatoxin B_1 metabolism. *Toxicology and Applied Pharmacology*, 2008, 227, 339-346.

[184] Murthy, PS; Ramalakshmi, K; Srinivas, P. Fungitoxic activity of Indian borage (*Plectranthus amboinicus*) volatiles. *Food Chemistry*, 2009, 114, 1014-1018.

[185] Boutigny, AL; Barreau, C; Atanasova-Penichon, V; Verdal-Bonnin, MN; Gadais, LP; Forget, FR. Ferulic acid, an efficient inhibitor of type B trichothecene biosynthesis and *Tri* gene expression in *Fusarium* liquid cultures. *Mycological Research*, 2009, 113, 746-753.

[186] Nesci A; Etcheverry M. Effect of natural maize phytochemicals on *Aspergillus* section *Flavi* sclerotia characteristics under different conditions of growth media and water potential. *Fungal Ecology*, 2009, 2, 44-51.*1*.

[187] Tsao, SM; Yin, MC. In vitro antimicrobial activity of four diallyl sulphides occurring naturally in garlic and Chinese leek oils. *Journal of Medical Microbiology*, 2001, 50, 646–649.

[188] Shukla, R; Kumar, A; Prasad, CS; Srivastava, S; Dubey, NK. Antimycotic and antiaflatoxigenic potency of *Adenocalymma alliaceum* Miers.on fungi causing biodeterioration of food commodities and raw herbal drugs. *International Biodeterioration & Biodegradation*, 2008, 62, 348-351.

[189] Rizzo, I; Vedoya, G; Maurutto, S; Haidukowski, M; Varsavsky, E. Assessment of toxigenic fungi on Argentinean medicinal herbs. *Microbiological Research*, 2004, 159, 113–120.

[190] Soliman, KM; Badeaa, RI. Effect of oil extracted from some medicinal plants on different mycotoxigenic fungi. *Food and Chemical Toxicology*, 2002, 40, 1669-1675.

[191] Bluma R; Amaiden MR; Etcheverry M. Screening of Argentine plant extracts: Impact on growth parameters and aflatoxin B_1 accumulation by *Aspergillus* section *Flavi*. *International Journal of Food Microbiology*, 2008, 122, 114-125.

[192] Singh P; Bhagat S; Ahmad SK. Aflatoxin elaboration and deterioration in some pulse cultivars during infestation with *Aspergillus flavus*. *Journal of Food Science and Technology,* 1990, 27, 60–62.

[193] Kumar A; Shukla R; Singh P; Prasad CS; Dubey NK. Biodeterioration of some herbal raw materials by storage fungi and aflatoxin and assessment of *Cymbopogon flexuosus* essential oil and its components as antifungal. *International, Biodeterioration & Biodegradation*, 2009, 63, 712-716.

[194] Chinaphuti, A; Aukkasarakul, S. Inhibition of *Aspergillus flavus* growth and aflatoxin production using herbal extracts. *Journal of Biotechnology*, 2008, *136, S752.*

[195] Razzaghi-Abyaneh, M; Shams-Ghahfarokhi, M; Yoshinari, T; Rezaee, M; Jaimand, K; Nagasawa, H; Sakuda, S; Tzortzakis, NG; Economakis, CD. Antifungal activity of lemongrass (*Cympopogon citratus* L.) essential oil against key postharvest pathogens. *Innovative Food Science & Emerging Technologies*, 2007, 8, 253-258.

[196] Thanaboripat, D; Mongkontanawur, N; Suvathi, Y; Ruangrattanamatee, V. Inhibition of aflatoxin production and growth of *Aspergillus flavus* by citronella oil. *KMITLScience Journal*, 2004, 49, 1-8.

[197] Juglal, S; Govinden, R; Odhav, B. Spice oils for the control of co-occurring mycotoxin producing fungi. *Journal of Food Protection*, 2002, 65, 683–687.

[198] Bluma, R; Amaiden, MR; Etcheverry, M. Screening of Argentine plant extracts: Impact on growth parameters and aflatoxin B_1 accumulation by *Aspergillus* section *Flavi*. *International Journal of Food Microbiology*, 2008, 122, 114-125.

[199] Sidhu, OP; Chandra, H; Behl, HM. Occurrence of aflatoxins in mahua (*Madhuca indica* Gmel.) seeds: Synergistic effect of plant extracts on inhibition of *Aspergillus flavus* growth and aflatoxin production. *Food and Chemical Toxicology*, 2009, 47, 774-777.

[200] Paster, N; Menasherov, M; Ravid, U; Juven, B. Antifungal activity of oregano and thyme essential oils applied as fumigants against fungi attacking stored grain. *Journal Food Protection*, 1995, 58, 81–85.

[201] Novak, J; Christina, B; Langbehn, B; Pank, F; Skoula, M; Gotsiou Y. Ratios of *cis*- and *trans*-sabinene hydrate in *Origanum marjoram* L. and *Orignaum midrophyllum* (Bentham) Vogel. *Biochemical Systematics and Ecology*, 2000, 28, 697–704.

[202] Aligianis, N; Kalpoutzakis, E; Mitaku, S; Chinou, IB. Composition and antimicrobial activity of the essential oil from *Origanum* species. *Journal of Agricultural and Food Chemistry*, 2001, 49, 4168–4170.

[203] Celiktas, Y; Hames, O; Kocabas, EE; Bedir, E; Vardar, SF; Ozek, T; Baser, KHC. Antimicrobial activities of methanol extracts and essential oils of Rosmarinus officinalis, depending on location and seasonal variations. *Food Chemistry*, 2007, 100, 553–559.

[204] Rasooli, I; Rezaei, MB; Allameh, A. Growth inhibition and morphological alterations of *Aspergillus niger* by essential oils from Thy- mus eriocalyx and Thymus x-porlock. *Food Control*, 2006, 17, 359–364.

INDEX

C

D

E

F

G

H

I

N

O

P

R

S